Engineering Dynamics

Engineering Dynamics

Fundamentals and Applications

M Rashad Islam

A K M Monayem H Mazumder

Mahbub Ahmed

CRC Press

Taylor & Francis Group

Boca Raton London New York

CRC Press is an imprint of the
Taylor & Francis Group, an **informa** business

First edition published 2022
by CRC Press
6000 Broken Sound Parkway NW, Suite 300, Boca Raton, FL 33487-2742

and by CRC Press
2 Park Square, Milton Park, Abingdon, Oxon, OX14 4RN

ISBN: 9781032255576 (hbk)
ISBN: 9781032255613 (pbk)
ISBN: 9781003283959 (ebk)

DOI: 10.1201/9781003283959

Typeset in Times
by Deanta Global Publishing Services, Chennai, India

Access the [companion website/Support Material]: www.routledge.com/9781032255576

Contents

Preface

This textbook is intended to be for the first class of engineering dynamics for undergraduate students in engineering and its related programs. Engineering dynamics is a rigorous topic in engineering with intensive use of vector mathematics and calculus. This textbook adopted plain language with an optimum amount of vector mathematics and calculus to introduce these topics to engineering students with some physics background. Numerous real-world examples are also provided with their step-by-step solutions. This is done keeping in mind the interests of new engineering and applied engineering students. The topics covered in the Fundamentals of Engineering (FE) examination are especially discussed in this textbook. A chapter on roadway dynamics has been included to make a bridge between engineering dynamics and transportation engineering.

The book has been thoroughly inspected with the help of professional editors to fix typos, editorial issues, and poor sentence structure. Despite this, if there is any issue, please excuse us and report it to the author at islamunm@gmail.com. The author is also requesting the readers to report any concern about the book. Any suggestion to improve this book, or any issue reported, will be fixed in the next edition with proper acknowledgement. Thanking you.

The Authors
islamunm@gmail.com

Authors

Dr. M Rashad Islam is a professor at the Colorado State University Pueblo. He is also a registered Professional Engineer (PE) and an ABET Program Evaluator. Dr. Islam received his PhD in civil engineering (with distinction in research, and GPA 4.0 out of 4.0 in coursework) from the University of New Mexico. He has an MS degree in civil engineering jointly from the University of Minho (Portugal) and Technical University of Catalonia (Barcelona, Spain). Dr. Islam has over 100 publications including text books, scholarly articles in top ranked journals, book chapters, and conference papers. His major textbooks include *Pavement Design – Materials, Analysis and Highways* (McGraw Hill), *Civil Engineering Materials – Introduction and Laboratory Testing* (CRC Press), *Engineering Statics* (CRC Press), and *Construction Safety – Health, Practices, and OSHA* (McGraw Hill). Dr. Islam also published a lot of PE civil engineering, FE civil engineering, and FE mechanical engineering exam preparation books using Amazon platform. Dr. Islam can be reached at islamunm@gmail.com.

Dr. A K M Monayem H Mazumder is an Associate Professor of Mechanical Engineering at the Saginaw Valley State University, Michigan. He earned his PhD in mechanical engineering from the University of Oklahoma and MS. in mechanical engineering from the University of New Orleans, Louisiana. Before joining the Saginaw Valley State University, he taught at Lamar University and Texas A&M University-Kingsville. Dr. Mazumder has several publications both in journals and conferences.

Dr. Mahbub Ahmed is an Associate Professor of Engineering at Southern Arkansas University. He earned his PhD with an emphasis in mechanical engineering from the University of Texas at El Paso and earned his master's in industrial engineering from Lamar University. He taught at Georgia Southern University from 2008 to 2012 in the Department of Mechanical Engineering. Dr. Ahmed teaches a wide variety of courses in engineering, engineering physics, engineering technology, and industrial technology. Dr. Ahmed also does consulting work for local industries. Dr. Ahmed currently holds a PE license in mechanical engineering in the state of Arkansas. He is also an ABET program evaluator for Engineering Technology and Accreditation Commission (ETAC).

1 Introduction

1.1 MECHANICS

Mechanics is a branch of science dealing with the action of force on a body at rest or in motion. Upon applying force on a body, two consequences may occur. First, it may move if there is not adequate restraint. Second, it may not move and is expected to deform. If the deformation is noticeable, then it is discussed under deformable-body mechanics (very often called Strength of Materials). If the deformation is negligible or movement occurs, then it is discussed under rigid-body mechanics. Both the rigid-body mechanics and deformable-body mechanics have two components:

1. Movement with constant speed or no movement
2. Movement with non-uniform velocity

If the body does not move or moves with constant speed, then it is discussed under Statics. If the body moves with non-uniform velocity, then the mechanism is discussed under Dynamics. If the body that is being discussed is liquid or gas, then the mechanics is discussed under Fluid Mechanics (Figure 1.1).

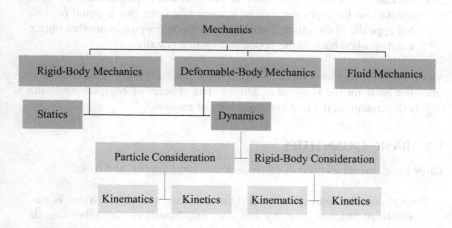

FIGURE 1.1 Branches of Mechanics

Upon applying load, the deformation is zero or so small it can be neglected in a rigid body. The distance between any two given points on a rigid body remains constant in time regardless of external forces or moments exerted on it. A particle is a rigid body, whose size can be ignored but mass cannot. The study of

DOI: 10.1201/9781003283959-1

1

describing movement (e.g., displacement, time, velocity, etc.) is known as kinematics. Kinetics is the study of forces that cause motion (e.g., torque, gravity, friction, etc.).

1.2 PRINCIPAL OF MECHANICS

Engineering mechanics is formulated based on Newton's three laws of motion that are stated below:

First Law: An object at rest will remain at rest unless acted upon by an unbalanced force. An object in motion continues in motion with the same speed and in the same direction unless acted upon by an unbalanced force. This means that there is a natural tendency of objects to keep on doing what they're doing. All objects resist changes in their state of motion. In the absence of an unbalanced force, an object in motion will maintain this state of motion.

Second Law: The summation of all forces on an object is equal to the mass of that object multiplied by the acceleration of that object. Acceleration is produced when a force acts on a mass. The greater the mass of the object (being accelerated), the greater the amount of force needed to accelerate that object. Everyone unconsciously knows (the Second Law) that heavier objects require more force to move the same distance as compared to lighter objects.

Third Law: For every action there is an equal and opposite reaction. This means that for every force there is a reaction force that is equal in size, but opposite in direction. That whenever an object pushes another object it gets pushed back in the opposite direction equally.

An overview of Newton's three laws of motion indicates that the second law provides the basis for the study of dynamics. The concept of engineering statics is largely dependent on the first and third laws of motion.

1.3 BASIC QUANTITIES

Three basic quantities used in mechanics are: length, mass, and time.

Time means the span spent on certain activities, action, etc. Time is naturally an absolute quantity and does not depend on any other outside object and proceeds uniformly at a fixed rate. Time interval between two events is equal for all observers.

Length is the linear measurement of an object. It is independent of observers, position, and time.

Mass is the quantity of matter in a body. It does not depend on the state of motion, place, and position.

1.4   BASICS OF UNITS

1.4.1   Types of Units

There are three types of units:

Fundamental Units: This unit does not depend on any other unit such as
unit of length or mass or time.

Derived Units: This unit is formed or derived from fundamental unit(s)
such as unit of force kg.m/s^2, unit of area m^2, etc.

Practical Units: Sometimes, fundamental, or derived units are so large or
so small that they are inconvenient for daily use. In those cases, sub-
multiples or multiples are used as units such as km, micron. Table 1.1
lists prefixes of some practical units.

TABLE 1.1
Prefixes

Multiple	Prefix	Symbol
10^{-18}	atto	a
10^{-15}	femto	f
10^{-12}	pico	p
10^{-9}	nano	n
10^{-6}	micro	μ
10^{-3}	milli	m
10^{-2}	centi	c
10^{-1}	deci	d
10^{1}	deka	da
10^{2}	hecto	h
10^{3}	kilo	k
10^{6}	mega	M
10^{9}	giga	G
10^{12}	tera	T
10^{15}	peta	P
10^{18}	exa	E

1.4.2   Systems of Units

There are several types of unit systems. Two are very popular and are discussed
here:

US Customary System (USCS) Units or FPS Units: In the US, length, mass, and time
are expressed in feet (ft), pound-mass (lbm) or slug, and second (sec), respectively.
FPS means foot-pound-second. Unit of mass is also used as slug (pound-mass),
which is pound divided by the gravitational constant (g = 32.2 ft/sec^2). More clearly,

$$\text{slug or lbm} = \frac{\text{lbf}}{\dfrac{\text{ft}}{\text{sec}^2}}$$

Unit of force is used as lbf or lb. Therefore, one must distinguish the pound-force (lbf) from the pound-mass (lbm). The pound-force is that force which accelerates one pound-mass at 32.2 ft/sec². Thus, 1 lbf = 32.2 lbm.ft/sec². In summary, lbf is written as lb and lbm is written as slug.

System of International (SI) Units: To maintain consistency worldwide SI units are proposed. In this system, length, mass, and time are expressed in meter (m), kilogram (kg), and second (sec), respectively. The conversion factors for major parameters used in dynamics from SI units to USCS units are listed in Table 1.2.

TABLE 1.2
Relationship between SI Units and USCS Units

Quantity	SI units	Multiply by	To obtain USCS units
Length	m	3.281	ft
Length	km	0.6214	mile
Mass	kg	2.205	lbf or lb
Mass	kg	0.069	lbm or slug
Force	N or kg.m/sec²	0.225	lbf or lb
Work	J or N.m	0.7376	ft.lbf (ft.lb)
Power	W	0.001341	horsepower (hp)

In this textbook, feet is written as ft; inch is written as in.; minute is written as min.; and second is written as sec. All other units are used to follow the most common practice.

Example 1.1

Convert 80 miles per hour (mph) into feet per second (fps).

SOLUTION

1 mile = 1/0.6214 km
= 1.6093 km
= 1.6093 × 1000 m
= 1609.3 × 3.281 ft
= 5,280 ft

Therefore, $80\,\dfrac{\text{mile}}{\text{hour}} = \dfrac{80\,\text{mile}\left(5,280\,\dfrac{\text{ft}}{\text{mile}}\right)}{(1\,\text{hour})3,600\,\dfrac{\text{sec}}{\text{hour}}} = 117\,\dfrac{\text{ft}}{\text{sec}}$

Answer: 117 fps

Note: feet per sec is very often written as fps.

1.5   ROUNDING OFF

Rounding off is a topic very confusing to every student, even to professionals. There are many customary rules for rounding off a number.

- If the last digit (the digit to be removed) is greater than 5, round up
- If the last digit (the digit to be removed) is smaller than 5, round down
- If the last digit (the digit to be removed) is 5:
  - If the digit preceding the 5 is an even number, then the digit is not rounded up
  - If the digit preceding the 5 is an odd number, then the digit is rounded up

Now, how many digits should be rounded? It depends on the number value, application, etc. To better understand, consider the weight of gold versus the weight of a vehicle. We try to express the weight of gold in three to four digits after the decimal if the unit is gram or oz. For kilogram or lb, the decimal places may be even larger. For example, 1.4123 g of gold, 0.354267 lb of gold. However, the weight of a car is commonly expressed as 3,200 lb or 3,250 lb. However, expressing the weight of a car as 3,237.25 lb looks non-practical. Another example is your age. If someone asks you your age, you commonly answer as 21 or 23 years or so. You do not answer as 21 years 5 months 3 days 12 hours or so. However, to determine the seniority by age, if two persons are 21 years, then we many need to go month or day or even to hour. The summary is that rounding off to some places depends on the value of the number and its application. Similarly, in mechanics, very often after the computation, the question of rounding off occurs. Students are advised to think logically. In this text, the authors round off numbers based on logic and customs.

In engineering analysis, where there are many steps of computation, rounding is performed at the final stage of the calculation. Otherwise, rounding in every step of the calculation may overestimate or underestimate the final results. For example, consider your test score in a class. Say, you made an average score of 7.7 (out of 10) in ten short quizzes that your instructor gave you. Your total quiz score is $7.7 \times 10 = 77$ out of 100. However, if your instructor rounds off the average quiz score as 8.0, then the total score is $8 \times 10 = 80$ which is pretty off from the actual total score due to rounding.

Example 1.2

Round off the following numbers using two decimal places.

(a)   287.365
(b)   287.364
(c)   287.367
(d)   287.315

SOLUTION

(a) 287.36 (If the digit preceding the 5 (i.e., 6) is an even number, then the digit is not rounded up. The digit 6 will not change)

(b) 287.36 (If the digit to be removed is smaller than 5, round down. The preceding digit, 6, will not change)

(c) 287.37 (If the digit to be removed is greater than 5, round up. The preceding digit, 6, will round up to 7)

(d) 287.32 (If the digit preceding the 5 (i.e., 1) is an odd number, then the digit is rounded up. The digit 1 will round up to 2)

FUNDAMENTALS OF ENGINEERING (FE) EXAM STYLE QUESTIONS

FE Problem 1.1

Say, you purchased 5 lb of sugar from Walmart. This 5 lb is the:

A. Mass
B. Weight
C. Slug
D. lbf times the gravitational acceleration

FE Problem 1.2

45 miles per hour (mph) equals most nearly:

A. 66 fps
B. 72 fps
C. 60 fps
D. 88 fps

FE Problem 1.3

45 miles per hour (mph) equals most nearly:

A. 72.1 fps
B. 20.1 m/sec
C. 18.1 m/sec
D. 56.2 fps

FE Problem 1.4

1 m/sec equals most nearly:

A. 3.281 fps
B. 0.305 fps
C. 39.36 fps
D. 12 fps

FE Problem 1.5

If a force acts on a 1-kg body and produces 1 m/sec^2 of acceleration, the value of the force is most nearly:

    A. 1 N
    B. 1 lb
    C. 1 kg.m^2/sec^2
    D. Cannot be determined

FE Problem 1.6

2 miles equals most nearly:

    A. 2,220 m
    B. 8,450 ft
    C. 3,220 m
    D. 3,600 m

2 Kinematics of Particles

2.1 ANALYTICAL ANALYSIS OF PARTICLE'S RECTILINEAR MOTION IN SURFACE

Kinematics is the study of the motion of particles, and rigid bodies, disregarding the forces associated with these motions. In other words, kinematics is the geometry of motion without considering the phenomenon that causes the motion. The motion can occur on a linear path or on a curved path. This section discusses the motion on a linear path. Rectilinear motion refers a linear motion in which the direction of the velocity remains constant and the path is a straight line.

Distance is defined as how far a particle travels from a reference point, while displacement is defined as how far a particle is from its initial position. Distance is a scalar quantity, while displacement is a vector quantity. More clearly, the complete length of the path between any two points is called distance. The distance can be determined by the total path the moving entity covers from the starting point regardless of its direction. Therefore, to calculate distance, the direction is not considered. Thus, the distance can only have positive values. In contrast, displacement is the direct length between any two points when measured along the minimum path between them. To calculate displacement, the direction must be taken into consideration. Thus, displacement is a vector quantity since both magnitude and direction are considered to determine it. Displacement can be positive, negative, and even zero.

The velocity of a particle can be determined by dividing the total displacement by the total time taken. This is the average velocity. However, the motion of the particle with time may be uniform or non-uniform. If the motion of a particle is non-uniform, the moving path can be divided into several elements. Velocity can be determined by dividing each of these displacement elements, ds, by the corresponding time elements, dt. An example is shown below in Figure 2.1.

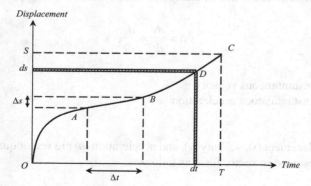

FIGURE 2.1 Distance–Time Variation of an Arbitrary Particle

DOI: 10.1201/9781003283959-2

The travel displacement of a particle with respect to time can be assumed as the *OABC* path. It travels a total of *S* displacement in *T* time. However, the motion of this particle is not linear in a segment of time (Δt) as it travels (Δs) from point *A* to point *B*.

If we consider the average velocity of the particle, then the average velocity can be expressed as:

$$v(avg.) = \frac{S}{T}$$

The average velocity of the segment *AB* can be expressed as:

$$v(avg. \text{ from } A \text{ to } B) = \frac{\Delta s}{\Delta t}$$

Using very small displacement in a very small time span, for example at point *D*, the velocity can be expressed as follows:

$$v = \frac{ds}{dt}$$

This velocity is also called instantaneous velocity. Some authors express this instantaneous velocity as follows:

$$v = \lim_{\Delta t \to 0} \frac{\Delta s}{\Delta t} = \frac{ds}{dt}$$

As Δt continues to decrease, the average velocity $\frac{\Delta s}{\Delta t}$ reaches a constant or limit-ing value. That means the instantaneous velocity at time *t* is the average velocity during the time interval Δt which approaches zero. It is also the slope of the dis-placement–time curve at time *t*.

Acceleration (*a*) is the rate of change of velocity, and can be expressed as:

$$a = \frac{dv}{dt} = \frac{d^2s}{dt^2}$$

where:
    v = The instantaneous velocity
    a = The instantaneous acceleration
    t = Time

The displacement (*s*), velocity (*v*), and acceleration (*a*) are vector quantities and can be represented in vector form as follows:

$$\vec{s} = x\vec{i} + y\vec{j} + z\vec{k}$$

$$\vec{v} = \dot{x}\vec{i} + \dot{y}\vec{j} + \dot{z}\vec{k}$$

$$\vec{a} = \ddot{x}\vec{i} + \ddot{y}\vec{j} + \ddot{z}\vec{k}$$

where x, y, and z are the components of displacements along x, y, and z axes, respectively. $\vec{i}, \vec{j}$ and $\vec{k}$ are the unit vectors along the x, y, and z axes, respectively.

$$\dot{x} = \frac{dx}{dt} = v_x$$

$$\ddot{x} = \frac{d^2x}{dt^2} = a_x$$

Rate of change of acceleration is not defined but considered as constant acceleration or variable acceleration. More clearly, the acceleration (or rate of change of velocity) may or may not be constant. If the acceleration is constant, some commonly used formulas used in kinematics can be derived. If the acceleration is not constant, the kinematics problems are solved using the calculus. For example, assume a particle is in motion with a constant acceleration, a_o, from initial time, t_o, to final time, t. The velocity of the particle at the initial time is v_o and at the final time is v. Then,

$$a_o = \frac{dv}{dt}$$

$$dv = a_o dt$$

$$\int_{v_o}^{v} dv = \int_{t_o}^{t} a_o dt$$

$$\left.v\right|_{v_o}^{v} = a_o \left.t\right|_{t_o}^{t}$$

$$v - v_o = a_o\left(t - t_o\right)$$

Therefore, $v = v_o + a_o\left(t - t_o\right)$

Following the above-mentioned procedure, the following equations can be derived:

$$s = s_o + v_o\left(t - t_o\right) + \frac{1}{2}a_o\left(t - t_o\right)^2$$

$$v^2 = v_o^2 + 2a_o\left(s - s_o\right)$$

In summary, the expressions for acceleration and velocity for uniform and non-uniform motion can be expressed as shown in Table 2.1.

TABLE 2.1

Summary of Equations for Rectilinear Motion

Variable, a	Constant, $a = a_o$
$a = \dfrac{dv}{dt}$	$v = v_o + a_o\left(t - t_o\right)$
$v = \dfrac{ds}{dt}$	$s = s_o + v_o\left(t - t_o\right) + \dfrac{1}{2}a_o\left(t - t_o\right)^2$
$ads = vdv$	$v^2 = v_o^2 + 2a_o\left(s - s_o\right)$

where:

s = Displacement along the line of travel
s_o = Displacement at time t_o
v = Velocity along the direction of travel
v_o = Velocity at time t_o
a_o = Constant acceleration
t = Time
t_o = Some initial time

If the initial point of the body is the reference point, i.e., $t_o = 0$ and $s_o = 0$, then the equations listed in Table 2.1 with constant acceleration (a_o) can be written as follows:

$$v = v_o + a_o t$$

$$s = v_o t + \frac{1}{2}a_o\left(t\right)^2$$

$$v^2 = v_o^2 + 2a_o s$$

Very often, velocity and speed are confused. Speed is defined as the rate at which an object travels along a path without considering the direction. In contrast, velocity is the rate and direction of an object's movement. Thus, speed is a scalar quantity, while velocity is a vector quantity. For constant acceleration, the displacement and velocity with respect to time consider can be shown in Figure 2.2.

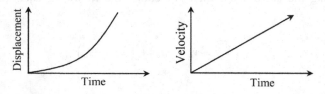

FIGURE 2.2 Displacement and Velocity with Time for Constant Acceleration

Instead of acceleration, if deceleration occurs, i.e., if the velocity decreases with time, then the acceleration (a) can be replaced with deceleration ($-a$). In this case, the equation will look like this:

$$v = v_o - a_o t$$

$$s = v_o t - \frac{1}{2} a_o (t)^2$$

$$v^2 = v_o^2 - 2a_o s$$

It is shown that some basic knowledge of calculus (differentiation and integration) is required to analyze dynamics problems. If you need to refresh your expertise on calculus, please review the calculus before proceeding. Now, practice the following worked-out examples to understand how the above equations are to be used while solving kinematics problems on particles. Remember the following steps to solve problems in dynamics.

Step 1. List the given parameters and required parameters with their appropriate symbols.

Step 2. Explore the equations available to find out the suitable one to use. Sometimes, a single equation may not be enough to reach the answer.

Step 3. Report the answer clearly with the appropriate unit.

Now consider an example of how these basic equations of distance, velocity, and acceleration can be used to solve kinematics problems without calculus.

Example 2.1

The velocity of a car increases from 5 m/sec to 20 m/sec in 10 sec. Determine the following:

(a) The acceleration of the car at the end of this 10-sec period.
(b) The displacement by the car during this 10-sec period.

SOLUTION

When beginning to solve this problem, list all the given parameters and the required parameters using their symbols. Here the given parameters are:

Initial velocity, $v_o = 5$ m/sec
Final velocity, $v = 20$ m/sec
Time, $t = 10$ sec
Acceleration after 10 sec, $a = ?$
Distance after 10 sec, $s = ?$

The next step is to explore the equations to find out which is the most suitable to use. Sometimes a single equation may not be enough to reach the answer. In this case, several equations, one after another may need to be used.

(a) The acceleration of the car at the end of 10 sec.
     [Think of which equation can be used to determine the accelera-
     tion after 10 sec. Given, the change in velocity, and the time, the only
     equation that can be used is $a = \dfrac{dv}{dt} = \dfrac{\Delta v}{\Delta t}$]

$$a = \frac{dv}{dt}$$

$$= \frac{\Delta v}{\Delta t}$$

$$= \frac{v - v_o}{t}$$

$$= \frac{20 \text{ m/sec} - 5 \text{ m/sec}}{10 \text{ sec}}$$

$$= \frac{15}{10} \frac{\text{m}}{\text{sec}^2}$$

$$= 1.5 \frac{\text{m}}{\text{sec}^2}$$

(b) The displacement by the car during the 10 sec.
     [While choosing equation(s) pay attention to the given data first.
     Then, find out which equation deals with these parameters]

$$s = v_o t + \frac{1}{2} a_o (t)^2$$

$$= (5 \text{ m/sec})(10 \text{ sec}) + \frac{1}{2}(1.5 \text{ m/sec}^2)(10 \text{ sec})^2$$

$$= 125 \text{ m}$$

Alternative: $v^2 = v_o^2 + 2a_o s$

$$s = \frac{v^2 - v_o^2}{2a_o}$$

$$= \frac{(20 \text{ m/sec})^2 - (5 \text{ m/sec})^2}{2(1.5 \text{ m/sec}^2)}$$

$$= 125 \text{ m}$$

Answers:

   1.5 m/sec²
   125 m

Now let us practice an example to see how these basic equations of displacement, velocity, and acceleration can be used to solve kinematics problems using calculus. Note that if the parameters vary with time non-uniformly, i.e., function of time, then calculus is used to solve it.

Example 2.2

A particle's displacement is represented by $s = 22t + t^2 + 0.5t^3$ m, where t is in sec. Determine the following:

(a) Particle's initial velocity
(b) Displacement in 5 sec
(c) Velocity after 5 sec
(d) Particle's initial acceleration
(e) Particle's acceleration after 2 sec.

SOLUTION

Given, $s = 22t + t^2 + 0.5t^3$

As the distance is a function of time, differentiation is required to determine velocity and acceleration.

(a) Velocity, $v = \dfrac{ds}{dt}$

$$= \frac{d}{dt}\left(22t + t^2 + 0.5t^3\right)$$

$$= 22 + 2t + 1.5t^2$$

When, $t = 0$, $v = 22$ m/sec

(b) Displacement, $s = 22t + t^2 + 0.5t^3$
when, $t = 5$ sec,

$$s = 22(5 \text{ sec}) + (5 \text{ sec})^2 + 0.5(5 \text{ sec})^3$$

$$= 197.5 \text{m}$$

(c) Velocity, $v = \dfrac{ds}{dt}$

$$= \frac{d}{dt}\left(22t + t^2 + 0.5t^3\right)$$

$$= 22 + 2t + 1.5t^2$$

When, $t = 5$ sec, $v = 22 + 2(5 \text{ sec}) + 1.5(5 \text{ sec})^2 = 69.5$ m/sec

(d) Velocity, $v = \dfrac{ds}{dt} = 22 + 2t + 1.5t^2$

$$a = \frac{dv}{dt}$$

$$= \frac{d}{dt}\left(22 + 2t + 1.5t^2\right)$$

$$= 2 + 3t$$

When, $t = 0$, $a = 2$ m/sec²

(e) $a = \frac{dv}{dt} = 2 + 3t$

When, $t = 2$ sec, $a = 2 + 3(2 \text{ sec}) = 8$ m/sec²

Answers:
22 m/sec
197.5 m
69.5 m/sec
2 m/sec²
8 m/sec²

Now work out the following problems with or without calculus. Different types of kinematics problems are shown here for a fuller understanding of a particle's kinematics.

Example 2.3

A vehicle traveling at 50 miles per hour (mph) saw the yellow signal when it was just approaching the junction. The driver of the vehicle knew that the yellow signal at this junction lasts for 4 sec and the junction is 150 ft long.

(a) Determine the vehicle's acceleration to cross the junction in time.
(b) Determine the vehicle's position after 4 sec if it travels accordingly.
(c) Determine the vehicle's velocity after 4 sec if it travels accordingly.

SOLUTION

(a) When beginning to solve this problem, list all the given parameters and required parameters using their symbols. The velocity is reported in miles per hour or mph, all other units are in the US system. Therefore, unit conversion is required first.

Initial velocity, $v_o = 50$ mph $= 50 \dfrac{\text{mile}}{\text{hr}} = 50 \dfrac{\text{mile}}{\text{hr}}\left(5,280 \dfrac{\text{ft}}{\text{mile}}\right)\left(\dfrac{1}{3,600} \dfrac{\text{hr}}{\text{sec}}\right)$
$= 73.33$ ft/sec

Time of travel, $t = 4$ sec
Displacement, $s = 150$ ft
Acceleration, $a_o = ?$

The next step is to explore the equations to find out which is the most suitable one to use. Sometimes a single equation may not be enough to reach the answer. Considering the above data, the equation of $s = v_o t + \dfrac{1}{2} a_o t^2$ can handle this as v_o, t, and s are known and a_o is to be found out.

Therefore, $s = v_o t + \dfrac{1}{2} a_o t^2$

$$150\,\text{ft} = \left(73.33\,\frac{\text{ft}}{\text{sec}}\right)(4\,\text{sec}) + \frac{1}{2}a(4\,\text{sec})^2$$

$$150\text{ft} = 293.33 + 8a_o$$

$a_o = -17.92\,\text{ft}/\text{sec}^2$ [negative sign means, even with 17.92 ft/sec² deceleration, the vehicle can still cross the junction. Therefore, no acceleration is required, and the vehicle can drive with its original velocity.]

(b) Vehicle's position after 4 sec
The car was running at constant speed of 73.33 ft/sec.

$$s = v_o t = \left(73.33\,\frac{\text{ft}}{\text{sec}}\right)(4\,\text{sec}) = 292.32\,\text{ft}$$

$292.32\,\text{ft} - 150\,\text{ft} = 143.32\,\text{ft}$ after crossing the junction.

(c) Vehicle's velocity after 4 sec
Vehicle is traveling at a constant speed; 73.33 ft/sec or 50 mph. Therefore, at any moment of traveling, its velocity is 73.33 ft/sec or 50 mph.

ANSWERS:

No acceleration is required; even 17.92 ft/sec² deceleration is good.
143.32 ft after crossing the junction.
73.33 ft/sec (50 mph)

Note: Miles per hour, kilometers per hour, and feet per second are written as mph, kmph, and fps, respectively. These abbreviations are used throughout this textbook.

Example 2.4

A truck weighing 5 tons is running at a speed of 36 kilometers per hour (kmph).

(a) To stop the truck in 10 m, how much deceleration is to be applied?
(b) To stop the truck in 10 m, how much time will be required?

SOLUTION

Initial velocity, $v_o = 36\,\dfrac{km}{hr} = 36\,\dfrac{km}{hr}\left(1{,}000\,\dfrac{m}{km}\right)\left(\dfrac{1}{3{,}600}\,\dfrac{hr}{sec}\right) = 10\ m/sec$

Distance, $s = 10\ m$

(a) *How much deceleration (a_o) is to be applied?*

The truck will stop at $v = 0$

Now, explore which equations deal with v, v_o, a_o, and s. The most suitable equation is $v^2 = v_o^2 - 2a_o s$.

$$v^2 = v_o^2 - 2a_o s$$

$$0 = \left(10\,\frac{m}{sec}\right)^2 - 2a_o(10\,m)$$

$$a_o = 5.0\,m/sec^2$$

(b) How much time (t) will be required?

Now, explore which equations deal with v, v_o, a_o, and t. The most suitable equation is $v = v_o - a_o t$.

$$v = v_o - a_o t$$

$$t = \frac{v_o - v}{a_o} = \frac{10\,m/sec - 0}{5.0\ m/sec^2} = 2.0\ sec$$

Answers:

Deceleration 5.0 m/sec²
Time 2 sec

Example 2.5

A body was at rest. A force of 15 N acted on it for 4 sec and then the force stopped. The body then moved 54 m in the next 9 sec.

(a) Determine the acceleration of the body when the force acted on it.
(b) Determine total displacement by the body.
(c) Determine velocity of the body at the end of the next 9 sec.

SOLUTION

(a) The force did not act after 4 sec. Therefore, in the next 9 sec, the body moved at a uniform velocity.

Therefore, $v = s/t = 54\text{ m}/9\text{ sec} = 6\text{ m/sec}$
This velocity (6 m/sec) the body attained when the force acted on it for 4 sec.
Therefore, $v = v_o + a_o t$

$$6 = 0 + a_o\left(4\sec\right)$$

$$a_o = 1.5\text{m}/\sec^2$$

(b) Displacement in the first 4 sec: $s = v_o t + \dfrac{1}{2}a_o t^2 = 0\left(4\sec\right) + \dfrac{1}{2}(1.5)\left(4\sec\right)^2$
 $= 12\text{ m}$
 Total displacement $= 12\text{ m} + 54\text{ m} = 66\text{ m}$
(c) The body was traveling at a constant speed for the last 9 sec. Therefore, at the end of the next 9 sec, the velocity was 6 m/sec.

Answers:

 Acceleration 1.5 m/sec²
 66 m
 6 m/sec

Example 2.6

A train starts from rest with an acceleration of 10 m/sec². Parallel to this train, a car starts at the same time with uniform speed of 100 m/sec. Determine when the train will overtake the car.

SOLUTION

Let us assume after t sec, the train will overtake the car.
 For train:

Initial velocity, $v_o = 0$
Acceleration, $a_o = 10\text{ m/sec}^2$
Displacement in t sec, $s = v_o t + \dfrac{1}{2}a_o t^2 = 0 + \dfrac{1}{2}(10\text{ m/sec}^2)(t^2)$

$$s = 5t^2$$

For car:

Uniform velocity, $v = 100\text{ m/sec}$
Displacement in t sec, $s = vt = 100t$
As they will overtake each other at t sec, the displacements in t sec will
 be equal.
Therefore, $100\,t = 5t^2$
$t(5t - 100) = 0$
$t = 0$ or 20 sec

Therefore, $t = 20$ sec

Answer: 20 sec

Example 2.7

A 10-kg block begins sliding down a 25° inclined plane. Starting from rest, the block attains a velocity of 5 m/sec after 5 sec. Determine how far the block will travel in 12 sec from the beginning.

SOLUTION

Given,

Mass, $m = 10$ kg
Initial velocity, $v_o = 0$
Velocity after 5 sec, $v = 5$ m/sec
Time, $t = 5$ sec
Displacement, $s = ?$

Starting from rest, the block attains a velocity of 5 m/sec after 5 sec due to gravitational force or any other force not mentioned. As the velocity is changing, there is some acceleration and we need to know the value of this acceleration to calculate the displacement.

Acceleration, $a_o = \dfrac{v - v_o}{t} = \dfrac{5\,\text{m/sec} - 0}{5\,\text{sec}} = 1.0\,\text{m/sec}^2$

Displacement in 5 sec, $s_1 = v_o t_1 + \dfrac{1}{2} a_o t_1^2$

$$s_1 = 0(5\,\text{sec}) + \dfrac{1}{2}(1\,\text{m/sec}^2)(5\,\text{sec})^2 = 12.5\,\text{m}$$

In the next 7 sec, the initial velocity will be 5 m/sec, as we are starting the calculation from the 6th second.

Displacement in the next 7 sec, $s_2 = v_o t_2 + \dfrac{1}{2} a_o t_2^2$

$$s_2 = (5\,\text{m/sec})(7\,\text{sec}) + \dfrac{1}{2}(1\,\text{m/sec}^2)(7\,\text{sec})^2 = 59.5\,\text{m}$$

Total displacement, $s = s_1 + s_2 = 12.5$ m + 59.5 m = 72 m
Alternative:
We can calculate the total displacement in 12 sec at a time. When we know the acceleration, initial velocity, and time, we can use the following equation.

$$s = v_o t + \dfrac{1}{2} a_o t^2 = 0(12\,\text{sec}) + \dfrac{1}{2}(1.0\,\text{m/sec}^2)(12\,\text{sec})^2 = 72\,\text{m}$$

Answer: 72 m

Example 2.8

Two engine-driven boats are racing with velocities of 10 m/sec and 5 m/sec. Their accelerations are 2 m/sec² and 3 m/sec², respectively. If the two boats reach the end at the same time, then, how long did those boats participate in the race? Also, calculate the racing distance.

SOLUTION

Let the racing time be t sec which is equal for both boats.
For the 1st boat:
Initial velocity, $v_o = 10$ m/sec
Acceleration, $a_o = 2$ m/sec²

Displacement by the first boat, $s_1 = v_{o1}t + \dfrac{1}{2}a_{o1}t^2$

$$= \left(10\,\frac{m}{sec}\right)t + \frac{1}{2}\left(2\,\frac{m}{sec^2}\right)(t^2)$$

$$= 10t + t^2$$

For the 2nd boat:
Initial velocity, $v_o = 5$ m/sec
Acceleration, $a_o = 3$ m/sec²

Displacement by the second boat, $s_2 = v_{o2}t + \dfrac{1}{2}a_{o2}t^2$

$$= \left(5\,\frac{m}{sec}\right)t + \frac{1}{2}\left(3\,\frac{m}{sec^2}\right)(t^2)$$

$$= 5t + 1.5t^2$$

As both boats reach the end at the same time, their displacements will be equal. Therefore,

$s_1 = s_2$
$10t + t^2 = 5t + 1.5t^2$
$0.5t^2 - 5\,t = 0$
$t(0.5t - 5) = 0$
$t = 0$ or 10 sec

Therefore, $t = 10$ sec

From the first equation: $s_1 = 10t + t^2 = 10(10\ sec) + (10\ sec)^2 = 200\,m$
You can use the second equation as well.
Answers:

10 sec
200 m

Example 2.9

A particle's velocity is represented by $v(t)=22 + 2t$ m/sec, where t is in sec.

  (a) Calculate how much the body will travel (displacement) in 10 sec.
  (b) Calculate its acceleration after 10 sec.

SOLUTION

As the velocity is given as a function of time, we need to use calculus.

  (a) Displacement in 10 sec

$$v = \frac{ds}{dt}$$

$$s = \int vdt = \int_{0}^{10} (22 + 2t)dt$$

$$= \left[22t + \frac{2}{2}t^2 \right]_{0}^{10}$$

$$= 320\,m$$

  (b) Acceleration after 10 sec

$$a = \frac{dv}{dt}$$

$$= \frac{d}{dt}(22 + 2t)$$

$$= 0 + 2$$

$$= 2$$

Acceleration is constant, 2 m/sec²

> *Answers:*
> *320 m*
> *2 m/sec²*

Example 2.10

An airplane moving with a velocity of 50 km/h touches a straight highway horizontally and comes to rest after running 300 m. If the deceleration (retardation) is uniform, calculate the time required to stop.

SOLUTION

Given data,

Initial velocity, $v_o = 50 \dfrac{km}{h} = \left(50 \dfrac{km}{h}\right) \dfrac{\left(1{,}000 \dfrac{m}{km}\right)}{3{,}600 \dfrac{sec}{h}} = 13.88 \dfrac{m}{sec}$

Final velocity, $v = 0$ (as the airplane stops)
Time to stop, $t = ?$
Displacement before stopping, $s = 300$ m

Now, go back to the equations to find out which equation deals with v, v_o, t, and s. There is no equation which deals with only these 4 parameters. If we use $v^2 = v_o^2 - 2a_o s$ and find out a_o, then we can use $v = v_o - a_o t$ to calculate t.

$v^2 = v_o^2 - 2a_o s$ [negative sign is used as acceleration is negative for deceleration]

$0 = (13.88 \text{ m/sec})^2 - 2a_o (300 \text{ m})$
$a_o = 0.321 \text{ m/sec}^2$

Again, $v = v_o - a_o t$

$0 = 13.88 - 0.321\, t$
$t = 42.9$ sec
$t \approx 43$ sec

Answer:

   43 sec

Example 2.11

A train traveling at 60 km/h is stopped at the next station in 10 sec by applying the brakes. Calculate the displacement from the next station where the brakes were applied.

SOLUTION

Given data,

Initial velocity, $v_o = 60 \dfrac{km}{h} = \left(60 \dfrac{km}{h}\right) \dfrac{\left(1{,}000 \dfrac{m}{km}\right)}{3{,}600 \dfrac{sec}{h}} = 16.67 \dfrac{m}{sec}$

Final velocity, $v = 0$ (as the train stops)
Time to stop, $t = 10$ sec
Displacement before stopping, $s = ?$

Now, go back to the equations to find out which equation deals with v, v_o, t, and s. There is no equation which deals with only these 4 parameters. If we use $v = v_o - a_o t$ and find a_o, then we can use $v^2 = v_o^2 - 2a_o s$ to calculate s.

Now, $v = v_o - a_o t$
$0 = 16.67 \text{ m/sec} - a_o (10 \text{ sec})$
$a_o = 1.67 \text{ m/sec}^2$

$$v^2 = v_o^2 - 2a_o s$$

$0 = (16.67 \text{ m/sec})^2 - 2(1.67 \text{ m/sec}^2) (s)$
$s = 83.3 \text{ m}$

Answer: 83.3 m

Example 2.12

A particle moving along in a straight line with a uniform acceleration covers 72 cm displacement at 12th sec. If the acceleration of the particle is 6 cm/sec², calculate the initial velocity.

SOLUTION

Displacement at 12th sec means the displacement in a single second. It can be calculated as the displacement in 12 sec minus the displacement in 11 sec. Let the initial velocity be v_o.

Given, acceleration, $a_o = 6 \text{ cm/sec}^2$.

Displacement in 12 sec, $s_{12} = v_o t + \dfrac{1}{2} a_o t^2 = v_o(12 \text{ sec}) + \dfrac{1}{2} \left(6 \dfrac{\text{cm}}{\text{sec}^2} \right)(12 \text{ sec})^2$
$= 12 v_o + 432$

Displacement in 11 sec, $s_{11} = v_o t + \dfrac{1}{2} a_o t^2 = v_o(11 \text{ sec}) + \dfrac{1}{2} \left(6 \dfrac{\text{cm}}{\text{sec}^2} \right)(11 \text{ sec})^2$
$= 11 v_o + 363$

Displacement at the 12th sec = $s_{12} - s_{11}$

$$72 \text{cm} = \left(12 v_o + 432 \right) - \left(11 v_o + 363 \right)$$

$$v_o = 3 \text{cm} / \text{sec}$$

Answer: 3 cm/sec

Sometimes, trigonometric rules are used to solve motions of two or more particles. If two concurrent motions act at a point, then the Parallelogram Method gives the magnitude and direction very quickly.

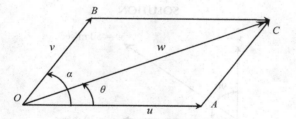

FIGURE 2.3 Demonstration of the Parallelogram Method

Let us assume two motions, u and v, are acting at O in a plane along OA and OB, respectively, with an angle of α, as shown in Figure 2.3. If a parallelogram $OACB$ can be drawn such that OC is the diagonal, then the diagonal OC can be represented as the resultant (w) of two motions u and v which are acting at O in a plane along OA and OB, respectively.

Then, the resultant (w) motion can be expressed as:

$$w = \sqrt{u^2 + v^2 + 2uv\cos\alpha}$$

If the angle between the motion (u) and the resultant (w) is θ with reference to u toward the v then it can be expressed as:

$$\theta = \tan^{-1}\left(\frac{v\sin\alpha}{u + v\cos\alpha}\right)$$

If the angle between the two motions is 90°, then $\cos 90 = 0$ and thus, $2uv\cos\alpha$ or $v\cos\alpha$ can be omitted and $\sin\alpha = 1$. Then, the equations will look like the Pythagoras equation for a right-angle triangle, as follows:

$$w = \sqrt{u^2 + v^2}$$

$$\theta = \tan^{-1}\left(\frac{v}{u}\right)$$

Now, let us look at an example of how the Parallelogram Method can be used to solve kinematics problems.

Example 2.13

A baseball is moving horizontally with a velocity of 30 m/sec which is hit with a bat at a right angle with the ball velocity, as shown in Figure 2.4. As a result, the ball acquires a velocity of 50 m/sec. Calculate the batting velocity.

SOLUTION

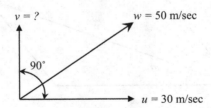

FIGURE 2.4 Sketch of Forces for Example 2.13

The ball attains the final velocity of 50 m/sec after the action of batting on the ball's initial velocity of 30 m/sec. The batting velocity and the ball's initial velocity produces the final velocity.

Ball's initial velocity, $u = 30$ m/sec
Ball's final velocity, $w = 50$ m/sec
Batting velocity, $v = ?$
Angle between u and v, $\alpha = 90$
Using the Parallelogram Method, $w^2 = u^2 + v^2 + 2uv \cos 90$

$$(50\,m/\sec)^2 = (30\,m/\sec)^2 + v^2 + 2(30\,m/\sec)v \cos 90$$

$$v^2 = (50\,m/\sec)^2 - (30\,m/\sec)^2$$

$$v = 40\,m/\sec$$

Answer: 40 m/sec

Example 2.14

A man swims at right angles to the current of a river and reaches the opposite bank 500 m from the starting point, as shown in Figure 2.5. If he swims at twice speed of the current's speed, calculate the width of the river.

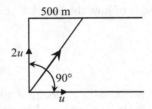

FIGURE 2.5 Sketch of Forces for Example 2.14

SOLUTION

The velocity of the river and the velocity of the swimmer are constant.
Therefore, $s = vt$ equation is valid both along the river and across the
river.
The velocity of the current is u and the swimming velocity is $2u$.
Displacement along the river: $s = ut$
Therefore, $t = 500/u$
Displacement across the river: $s = 2ut$
$s = 2u (500/u)$
$s = 1,000$ m
Answer: 1,000 m

Example 2.15

A man can swim directly across a river of 100 m width in 4 min if there is
no current. He takes 5 min to cross the river directly when there is current.
Calculate the velocity of the current.

SOLUTION

Let,

  u = river's current velocity
  v = swimmer velocity

If there is no current, $s = vt$

  $v = s/t = 100$ m $/ 4$ min $= 25$ m/min

If there is current, the following condition will occur.

  OB = swimmer velocity
  $OC = w$ = resultant velocity
  OD = river's current velocity

Let us draw a parallel line CD with OB; $OB = DC = v$ = swimmer velocity, as
shown in Figure 2.6.

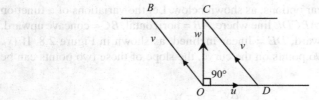

FIGURE 2.6   Sketch of Forces for Example 2.15

Again, $s = wt$

$w = 100 \text{ m} / 5 \text{ min} = 20 \text{ m/min}$

The resultant velocity will be perpendicular to the river velocity.
Therefore, from the Pythagoras equation, $v^2 = u^2 + w^2$

$$u = \sqrt{v^2 - w^2} = \sqrt{\left(25\frac{m}{min}\right)^2 - \left(20\frac{m}{min}\right)^2} = 15\frac{m}{min}$$

Answer: 15 m/min

The above-discussed section shows how kinematics problems can be solved using equations or calculus. In the next section, another procedure to solve kinematics problems using the graphical method is discussed.

2.2 GRAPHICAL ANALYSIS OF PARTICLE'S RECTILINEAR MOTION

2.2.1 GRAPHICAL SYSTEM

Cartesian coordinate systems (x, y, and z) are used for graphical presentation of most functions commonly used in engineering. The general form of straight-line equation is expressed as, $y = mx + c$, and can be figured, as shown in Figure 2.7:

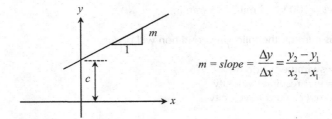

$$m = slope = \frac{\Delta y}{\Delta x} = \frac{y_2 - y_1}{x_2 - x_1}$$

FIGURE 2.7   Slope of a Straight Line in the Cartesian Coordinate System

Non-linear lines can be concave upward, concave downward, or a combination of both including linear options, as shown below. Let the variations of a function be represented by the $ABCDE$ line where, AB = horizontal, BC = concave upward, CD = concave downward, DE = linear inclined, as shown in Figure 2.8. If (x_1, y_1), and (x_2, y_2) are two points on the curve, the slope of these two points can be written as:

$$slope = \frac{\Delta y}{\Delta x} = \frac{y_2 - y_1}{x_2 - x_1} = \frac{rise}{run}$$

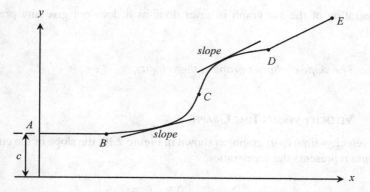

FIGURE 2.8 Slope of a Curved Line in the Cartesian Coordinate System

Thus, the slopes of different regions are listed below:

AB: Slope is zero as the rise is zero for a unit run
BC: Positive increasing as the rise increases with the run
CD: Positive decreasing as the rise decreases with the run
DE: Positive constant as the rise is constant for a unit run

2.2.2 Displacement versus Time Graph

In the position–time (*s–t*) graph, the slope of the curve at any point represents the velocity:

$$v = \frac{ds}{dt} = \frac{\Delta s}{\Delta t} = \frac{s_2 - s_1}{t_2 - t_1}$$

Therefore, the slope of the *s–t* graph can be determined more accurately by differentiating, or less accurately when considering a small segment, as shown in Figure 2.9.

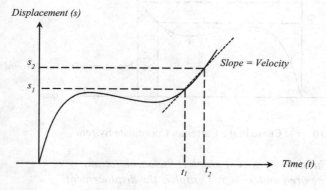

FIGURE 2.9 *s–t* Curve in the Cartesian Coordinate System

Integration of the s–t graph is never done as it does not give any practical quantity.

Rule: The slope of the s–t graph is the velocity.

2.2.3 VELOCITY VERSUS TIME GRAPH

In the velocity–time (v–t) graph, as shown in Figure 2.10, the slope of the curve at any point represents the acceleration:

$$a = \frac{dv}{dt} = \frac{\Delta v}{\Delta t} = \frac{v_2 - v_1}{t_2 - t_1}$$

The area of the s–t curve under certain limits represents the displacement during that period. Consider the following equations:

$$v = \frac{ds}{dt}$$

On integrating, $\int ds = \int v dt$

$$\int_{s=s_1}^{s=s_2} ds = \int_{t=t_1}^{t=t_2} v dt$$

This means the integration of v–t graph represents the displacement traveled for the limits the integration is performed.

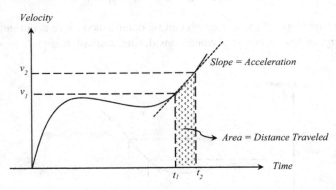

FIGURE 2.10 v–t Curve in the Cartesian Coordinate System

Rule: The slope of the v–t graph is the acceleration.
Rule: The area under the v–t graph is the displacement.

2.2.4 Acceleration versus Time Graph

The acceleration–time (a–t) graph is rarely used in kinematics. The slope of the a–t curve at any point does not represent any practical quantity.

Consider the following equation:

$$a = \frac{dv}{dt}$$

On integrating, $\int dv = \int a\,dt$

$$\int_{v=v_1}^{v=v_2} dv = \int_{t=t_1}^{t=t_2} a\,dt$$

This means the integration of a–t graph represents the change in velocity for the limits the integration is performed, as shown in Figure 2.11.

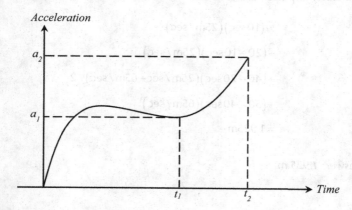

FIGURE 2.11 a–t Curve in the Cartesian Coordinate System

> *Rule: Area under the a–t graph is the change in velocity during the time range, or v at any point is the area under the a–t curve up to the point.*

Example 2.16

The velocity distribution of a hypothetical speed train is shown in Figure 2.12. The vertical axis shows the velocity in m/sec and the horizontal axis represents the time in sec. Calculate the displacement in 50 sec.

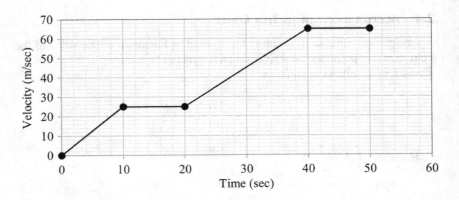

FIGURE 2.12 The v–t Curve for Example 2.16

SOLUTION

Displacement is calculated as the area under the v–t curve.
   Therefore, displacement in 50 sec:

$$\tfrac{1}{2}(10\,\text{sec})(25\text{m}/\text{sec})$$

$$+(20-10\,\text{sec})(25\text{m}/\text{sec})$$

$$+(40-20\,\text{sec})(25\text{m}/\text{sec}+65\text{m}/\text{sec})/2$$

$$+(50-40\,\text{sec})(65\text{m}/\text{sec})$$

$$=1{,}925\text{m}$$

Answer: 1,925 m

Example 2.17

The displacement–time (s–t) graph of a car is shown in Figure 2.13. The vertical axis shows the position in feet and the horizontal axis represents the time
in sec.
   Determine the following:

   (a) The velocity of the car at 2 sec
   (b) The velocity of the car at 14 sec
   (c) The acceleration of the car at 10 sec
   (d) The acceleration of the car at 6 sec
   (e) Total displacement in 8 sec
   (f) Total displacement in 20 sec
   (g) Total distance in 20 sec.

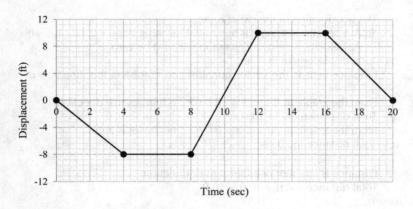

FIGURE 2.13 The s–t Curve for Example 2.17

SOLUTION

(a) The velocity of the car at 2 sec.

We know that the slope of the s–t graph is the velocity, i.e.,

$$v = \frac{\Delta s}{\Delta t} = \frac{s_2 - s_1}{t_2 - t_1}$$

$$v_{2nd} = \frac{s_2 - s_1}{t_2 - t_1} = \frac{-4\,\text{ft} - 0\,\text{ft}}{2\,\text{sec} - 0\,\text{sec}} = -2\,\frac{\text{ft}}{\text{sec}}$$

(b) The velocity of the car at 14 sec.

At 14 sec, the s–t graph is horizontal meaning slope = 0.

$$v_{14th} = 0\,\frac{\text{ft}}{\text{sec}}$$

(c) The acceleration of the car at 10 sec.

Acceleration is the change of velocity over time, i.e.,

$$a = \frac{dv}{dt} = \frac{\Delta v}{\Delta t} = \frac{v_2 - v_1}{t_2 - t_1}$$

Slope of s–t graph is the velocity, i.e., $v = \frac{ds}{dt} = \frac{\Delta s}{\Delta t} = \frac{s_2 - s_1}{t_2 - t_1}$

Considering (12 sec, 10 ft) and (8 sec, −8 ft),

$$v_{10th} = \frac{s_2 - s_1}{t_2 - t_1} = \frac{10\,\text{ft} - (-8\,\text{ft})}{12\,\text{sec} - 8\,\text{sec}} = 4.5\,\frac{\text{ft}}{\text{sec}}$$

As the velocity at 10 sec is constant, the acceleration is zero.

Rule: Velocity is constant meaning acceleration is zero

(d) The acceleration of the car at 6 sec.
     At 6 sec, the s–t graph is horizontal meaning slope = 0, i.e., $v_{6th} = 0$
     As velocity is zero, acceleration will also be zero.
(e) Total displacement in 8 sec.
     Displacement is the vector quantity and considers the direction of
        travel.
     From the graph, at 8 sec, $s = -8$ ft (i.e., 8 ft in reverse direction)
(f) Total displacement in 20 sec.
     From the graph, at 20 sec, $s = 0$ ft (i.e., net travel is zero)
(g) Total distance in 20 sec

     Distance is ignorant of either sign or direction.
     Total distance = 8 ft + 10 ft = 10 ft
Answers:

   (a) *–2 ft/sec*
   (b) *0 ft/sec*
   (c) *Acceleration is zero*
   (d) *Acceleration is zero*
   (e) *–8 ft*
   (f) *0 ft*
   (g) *18 ft*

Example 2.18

The velocity–time (v–t) graph of a car is shown in Figure 2.14. The vertical axis
shows the velocity in feet per sec and the horizontal axis represents the time
in sec.

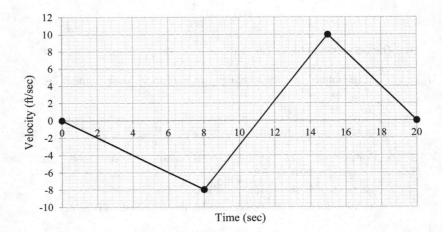

FIGURE 2.14   The v–t Curve for Example 2.18

Determine the following:

(a) The acceleration of the car at 6 sec
(b) Displacement in 15 sec
(c) The velocity at the 20 sec.

SOLUTION

(a) The acceleration of the car at 6 sec.

Acceleration is the slope of v–t graph, i.e., $a = \dfrac{dv}{dt} = \dfrac{\Delta v}{\Delta t} = \dfrac{v_2 - v_1}{t_2 - t_1}$

The slope of v–t graph is constant until 8th sec.
Considering (8 sec, −8 ft/sec) and (0 sec, 0 ft/sec),

$a_6 = \dfrac{v_2 - v_1}{t_2 - t_1} = \dfrac{-8\,\text{ft} - 0\,\text{ft}}{8\,\text{sec} - 0\,\text{sec}} = -1\,\dfrac{\text{ft}}{\text{sec}^2}$

(b) Displacement in 15 sec.
Area under v–t graph is the displacement.
Area under v–t graph from 0 to 12 sec = ½ (11 sec) (−8 ft/sec) + ½
(15 − 11 sec) (10 ft/sec) = −24 ft
(c) The velocity at 20 sec.

Velocity at 20 sec is zero as the graph shows.
Answers:

(a) *−1 ft/sec²*
(b) *−24 ft*
(c) *0 ft*

Example 2.19

The acceleration–time graph of a test car is shown in Figure 2.15. The vertical axis shows the acceleration in feet per sec² and the horizontal axis represents the time in sec.
    Determine the following:

(a) The velocity of the test car at 4 sec
(b) The velocity of the test car at 10 sec
(c) Displacement in 12 sec.

SOLUTION

(a) The velocity of the test car at 4 sec.
Area of a–t curve up to 4 sec = 10 ft/sec/sec (4 sec) = 40 ft/sec
(b) The velocity of the test car at 10 sec.
Area of a–t curve up to 10 sec = 10 ft/sec/sec (8 sec) + 4 ft/sec/sec (10
sec − 8 sec) = 88 ft/sec
Remember: v at any point is the area under the a–t curve up to that point

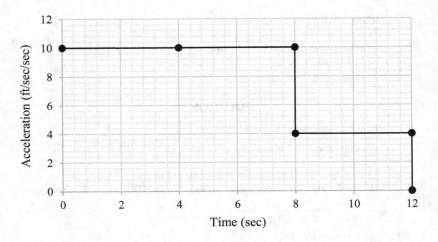

FIGURE 2.15 The v–t Curve for Example 2.19

(c) Displacement in 12 sec.

Displacement is the area of the v–t curve. Let us find the equation of v first. Considering the first part, 0 to 8 sec.

$$\int dv = \int a\,dt$$

$$\int_0^v dv = \int_0^t 10\,dt$$

$$|v|_0^v = |10t|_0^t$$

$$v - 0 = 10(t - 0)$$

$$v = 10t \quad [0 \leq t \leq 8 \text{ sec}]$$

$$v_8 = 10(8\,\text{sec}) = 80\,\frac{\text{ft}}{\text{sec}}$$

Considering the second part, 8 to 12 sec $\int dv = \int a\,dt$

$$\int_{v=v_8}^{v=v} dv = \int_{8\,\text{sec}}^t 4\,dt$$

$$|v|_{v_8}^v = |4t|_8^t$$

$$v - 80 = 4(t - 8)$$

$$v = 4t + 48 \ [8 \leq t \leq 12 \text{ sec}]$$

Displacement in 12 sec, $\displaystyle\int_{s=0}^{s=s} ds = \int_{t=0}^{t=8 \text{ sec}} v dt + \int_{t=8 \text{ sec}}^{t=12 \text{ sec}} v dt$

$$\int_{s=0}^{s=s} ds = \int_{t=0}^{t=8 \text{ sec}} 10t \, dt + \int_{t=8 \text{ sec}}^{t=12 \text{ sec}} (4t + 48) \, dt$$

$$[s]_0^s = \left[\frac{10t^2}{2}\right]_0^{8 \text{ sec}} + \left[\frac{4t^2}{2} + 48t\right]_{8 \text{ sec}}^{12 \text{ sec}}$$

$$s = \left[5(8^2 - 0^2) - 0\right] + \left[2(12^2 - 8^2) + 48(12 - 8)\right]$$

$$s = 320 + 352 \text{ ft}$$
$$s = 672 \text{ ft}$$

An alternative way to find displacement is to draw the v–t curve first. Area under the a–t curve is the velocity and then, the area under the v–t curve is the displacement.

Answers:

(a) 80 ft/sec
(b) 88 ft/sec
(c) 672 ft

2.3 MOTION OF FALLING PARTICLES

The previous sections discuss the particle's motion in a straight path on the surface of the Earth (say, the motion of a car on a flat and straight highway). The current section discusses the motion of a particle in a straight path but not on the surface, rather vertically straight upward or downward or a combination of both. What is the difference between these straight (rectilinear) motions on the surface and in vertical up/down? When a particle (say, a car) is in motion on a flat surface, gravity does not affect its motion as gravity works along the vertical direction. The motion on a flat surface follows Newton's rules as discussed in the previous sections. However, when a particle (say, a rocket) is in motion vertically upward with some velocity, the gravitational acceleration (g) acts downward and pulls the particle vertically downward, and the upward velocity decreases with time due to the downward gravitational pull. The upward velocity eventually becomes zero. Then, gravity pulls the particle downward and it starts to fall back. The point where the vertical component of the velocity is zero is the highest point the peak point; vertical velocity is zero at this point. During the return (falling back) the vertical downward velocity increases due to the action of gravitational acceleration. The velocity becomes at maximum just before touching the ground. Recall that in the whole travel path (upward and then downward), there is no horizontal velocity involved. Any velocity in upward/downward travel is the vertical velocity. Another topic to be discussed is the sign convention. If upward direction is assumed positive, then the downward direction is negative and vice-versa. Therefore, a stone thrown vertically up, travels upward, reaches its peak (say h) and falls back to the ground traveling the same distance (say, $-h$), as shown in Figure 2.16. The total displacement is zero because of $+h - h = 0$.

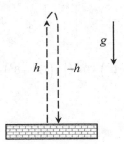

FIGURE 2.16 Travel Path for a Vertically Upward Thrown Particle

With computation, all the equations listed in the previous section for particles' rectilinear motion apply except acceleration (a_o) is replaced by gravitational acceleration (g). The value of g is considered constant at a certain point in the universe and acts downward. The distance traveled, s is commonly replaced by the vertical displacement, y or h. If a particle is let to fall downward at initial vertical downward velocity of v_{oy}, from a height of y, then the vertical downward velocity at any time, t is:

$$v_y = v_{oy} + gt$$

Where v_y, v_{oy}, and g work downward and all are assumed positive. If you assume the opposite sign, the results will be the same. Similarly, for falling particles, the other equations of motion can be written as:

$$y = v_{oy}t + \frac{1}{2}gt^2$$

$$v_y^2 = v_{oy}^2 + 2gy$$

For particles thrown upward at initial vertical upward velocity of v_{oy}, the equations of motion can be written as (considering upward positive):

$$v_y = v_{oy} - gt$$

$$y = v_{oy}t - \frac{1}{2}gt^2$$

$$v_y^2 = v_{oy}^2 - 2gy$$

The negative sign comes from the downward direction of gravitational acceleration (g) which is assumed negative here. Now, practice the following worked-out
problems to learn how the above equations are used. Remember, the gravitational
acceleration (g) always acts downward. The displacement is considered the algebraic sum. For example, if a particle travels upward by 5 feet (say, upward is positive) and returns down to the same point, the distance traveled is $5 - 5 = 0$ feet.

Remember: The value of gravitational acceleration (g) *is 9.81 m/sec² or
32.2 ft/sec².*

Example 2.20

A ball is thrown vertically upward from the top of a building with a velocity of
5 m/sec reaching the ground after 15 sec. Calculate the height of the building.

SOLUTION

After being thrown upward from the top of the building, the ball will reach its
peak, come back to the level of the top of building, and then go to ground.
Therefore, the effective travel path is from the top of building to the ground, as
shown in Figure 2.17. Let us assume that the downward direction is positive.

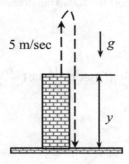

FIGURE 2.17 Sketch for Example 2.20

Initial velocity, v_o = 5 m/sec (upward, negative)
Gravitational acceleration, g = 9.81 m/sec² (downward, positive)
Time, t = 15 sec

Now, explore which available equation deals with y, v_o, t, and g. The equation $y = v_{oy}t + \frac{1}{2}gt^2$ is the most suitable. Therefore,

$$y = v_{oy}t + \frac{1}{2}gt^2$$

$$= \left(-5\frac{m}{sec}\right)(15\ sec) + \frac{1}{2}\left(9.81\frac{m}{sec^2}\right)(15\,sec)^2$$

$$= 1{,}029\ m$$

Answer: 1,029 m

Example 2.21

A balloon is ascending vertically with a velocity of 9 m/sec and a stone is let fall from it. If the stone reaches the ground in 10 sec, calculate the height of the balloon when the stone is let fall.

SOLUTION

After letting the stone fall, it will initially travel upward then return to its original releasing point, and then fall to the ground. Therefore, the effective travel path is from the releasing point to the ground. Let us assume that the downward direction is positive.

Initial velocity, v_o = 9 m/sec (upward, negative)
Gravitational acceleration, g = 9.81 m/sec² (downward, positive)
Time, t = 10 sec
Now, explore which available equation deals with y, v_o, t, and g. The equation $y = v_{oy}t + \frac{1}{2}gt^2$ is the most suitable. Therefore,

$$y = v_{oy}t + \frac{1}{2}gt^2$$

$$= \left(-9\frac{m}{sec}\right)(10\,sec) + \frac{1}{2}\left(9.81\frac{m}{sec^2}\right)(10\,sec)^2$$

$$= 400\ m$$

Answer: 400 m

2.4 PARTICLE'S PROJECTILE MOTION

So far, we have looked at straight-line motion (horizontal or vertical). Now, in this section, a non-straight-line motion is discussed. If a particle is thrown upward with some angle with respect to the horizon, then the particle's travel path is called the projectile motion. In a Projectile motion an object or a particle (a projectile) is thrown that moves along a curved path under the influence of gravity only (in particular, the effects of air resistance are assumed to be negligible). The curved path can be shown to be a parabola. The only force of significance that acts on the object is gravity, which always acts downward, thus imparting to the object a downward acceleration. Two points to remember:

- In the horizontal direction, there are no external forces. Therefore, there is no horizontal acceleration. Also, the velocity in the horizontal direction is constant.
- In the vertical direction, the only force acting on the projectile is gravity.

Let us consider a particle thrown at the initial velocity of v_o and an angle of θ with respect to the horizon, as shown in Figure 2.18. The horizontal component of the velocity ($v_x = v_o \cos\theta$) does not change with time, i.e., at any moment of the travel path, $v_x = v_o \cos\theta$. The vertical component of the velocity (v_y) changes with time as it is affected by the gravitational acceleration (g) which always acts downward. Therefore, the vertical component of the velocity (v_y) decreases with height and becomes zero when the particle reaches its peak. Then, the particle starts falling downward with the action of gravitational acceleration (g). The downward vertical velocity is highest when it just touches the ground. In the whole travel path, the horizontal component of the velocity remains the same, i.e., $v_x = v_o \cos\theta$.

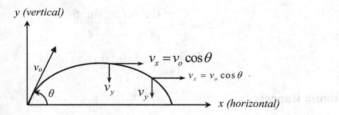

FIGURE 2.18 Travel Path of a Projectile

The initial horizontal component of the velocity is $v_x = v_o \cos\theta$. It always remains the same.

The acceleration is the action on the body given as $a_y = -g$. It is always vertically downward.

The equations for a common projectile motion may be obtained from the constant acceleration equation as:

Vertical acceleration, $a_y = -g$ (always acts downward)
Horizontal acceleration, $a_x = 0$ (always zero for any type of projectile)
Horizontal velocity, $v_x = v_o \cos\theta$ (always constant for projectiles)
Vertical velocity, $v_y = v_o \sin\theta - gt$ (decreases upward)
Horizontal displacement, $x = x_o + (v_o \cos\theta)t$

Vertical displacement, $y = y_o + (v_o \sin\theta)t - \dfrac{1}{2}gt^2$

Time to Reach the Peak:

Vertical velocity at the peak = 0. Therefore:

$$v_y = v_o \sin\theta - gt$$

$$0 = v_o \sin\theta - gt$$

Therefore, time to reach the peak is, $t = \dfrac{v_o \sin\theta}{g}$

Maximum Height:

If the particle starts from the ground, vertical distance traveled is

$y = (v_o \sin\theta)t - \dfrac{1}{2}gt^2$

Time to reach the maximum height or peak, $t = \dfrac{v_o \sin\theta}{g}$

The maximum height possible, $y_{max} = (v_o \sin\theta)\left(\dfrac{v_o \sin\theta}{g}\right) - \dfrac{1}{2}g\left(\dfrac{v_o \sin\theta}{g}\right)^2$

$$y_{max} = \dfrac{v_o^2 \sin^2\theta}{g} - \dfrac{v_o^2 \sin^2\theta}{2g}$$

$$y_{max} = \dfrac{v_o^2 \sin^2\theta}{2g}$$

Maximum Range:

Time from ground to reach the peak and time from the peak to return to ground are equal. Therefore:

Total travel time or time of flight, $T = 2t = \dfrac{2v_o \sin\theta}{g}$

Maximum horizontal distance traveled, $x_{max} = (v_o \cos\theta)T = (v_o \cos\theta)\dfrac{2v_o \sin\theta}{g}$

$$x_{max} = \dfrac{v_o^2(2\sin\theta\cos\theta)}{g} = \dfrac{v_o^2 \sin 2\theta}{g}$$

Trajectory:

The trajectory of a projectile can be obtained by eliminating the time variable t from the above equations and by solving for $y(x)$. We take $x_o = y_o = 0$ so the projectile is launched from the origin. The kinematic equation for x gives:

Horizontal displacement, $x = (v_o \cos\theta)t$

$$t = \frac{x}{v_o \cos\theta}$$

Vertical displacement, $y = (v_o \sin\theta)t - \frac{1}{2}gt^2$

$$y = (v_o \sin\theta)\left(\frac{x}{v_o \cos\theta}\right) - \frac{1}{2}g\left(\frac{x}{v_o \cos\theta}\right)^2$$

$$y = (\tan\theta)x - \frac{g}{2(v_o \cos\theta)^2}x^2$$

This trajectory equation, $y = ax + bx^2$, is an equation of a parabola with coefficients:

$$a = \tan\theta$$

$$b = -\frac{g}{2(v_o \cos\theta)^2}$$

The flow path of a projectile is thus a parabola. Some basic rules for projectile motion are listed below.

Rule: Vertical component of the velocity decreases with height, reaches zero at the peak, and increases downward when returning.

Rule: Gravitational acceleration always acts downward and is considered constant at a plane.

Rule: There is no horizontal acceleration for projectile motion.

Rule: There is no change in velocity along the horizontal direction.

Rule: At any point of the travel path, velocity has both x and y components.

Rule: The displacement is the algebraic sum. For example, if a stone is thrown upward from the ground and it falls to ground somewhere else, then the vertical distance traveled is zero.

Rule: While analyzing, the motion can be treated as two independent one-dimensional motions: One horizontal and the other vertical.

Example 2.22

A projectile with an initial velocity of 50 m/sec is launched at an angle of 45° to the horizon. Calculate the horizontal displacement of the projectile.

SOLUTION

Given,

Horizontal component of initial velocity, $v_{x,0} = v_o \cos\theta = \left(50\dfrac{m}{sec}\right)\cos 45$

Vertical component of initial velocity, $v_{oy} = v_o \sin\theta = \left(50\dfrac{m}{sec}\right)\sin 45$

Horizontal displacement, $x = x_o + (v_o \cos\theta)t = ?$

If the travel time is known, x can be determined.

For projectiles, vertical velocity at the peak point is always zero. Let the time to reach the peak be t_p.

Therefore, $v_y = -gt_p + v_o \sin\theta$

$0 = (-9.81 \text{ m/sec}^2)\, t_p + 50 \sin 45$

$t_p = 3.6 \text{ sec}$

This time (3.6 sec) is needed to reach the peak; an equal amount of time is required to return back to the ground. Therefore, total travel time $t = 2\, t_p = 2(3.6 \text{ sec}) = 7.2 \text{ sec}$.

Alternatively, you can calculate the total travel time or time of flight as

$\dfrac{2v_o \sin\theta}{g}$

Finally, horizontal displacement, $x = x_o + (v_o \cos\theta)t$

$x = 0 + 50 \cos 45\, (7.2 \text{ sec}) = 255 \text{ m}$

Answer: 255 m

Alternatively, you can use the equation, $x_{max} = \dfrac{v_o^2 \sin 2\theta}{g}$

Example 2.23

A cannon shell is fired with a velocity of 40 m/sec toward an enemy at an angle of 70° with respect to the horizon. Calculate the height the shell will strike the wall at 30 m away.

SOLUTION

As shown in Figure 2.19:

Horizontal component of initial velocity, $v_{ox} = v_o \cos 70 = (40\,\text{m/sec})$
$\cos 70 = 13.68 \text{ m/sec}$

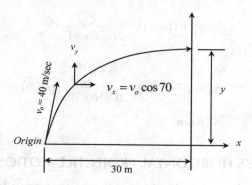

FIGURE 2.19 The Travel Path of the Projectile for Example 2.23

Vertical component of initial velocity, $v_{oy} = v_o \sin 70 = (40\,\text{m/sec})$
$\sin 70 = 37.59$ m/sec

Horizontal displacement, $x = 30$ m
Vertical height, $y = ?$

$$y = v_{oy}t - \frac{1}{2}gt^2$$

If the travel time is known, x can be determined.

Horizontal Travel: Recall that the horizontal component of velocity of a projectile is always constant, i.e., does not change with time. The vertical component changes due to the gravitational acceleration (g).

Therefore, $x = v_{ox}t$

30 m = 13.68 m/sec (t)
$t = 2.19$ sec

Vertical displacement after time, t: (assuming upward positive)

$$y = v_{oy}t - \frac{1}{2}gt^2$$

$$= (37.59\,\text{m/sec})(2.19\,\text{sec}) - \frac{1}{2}(9.81\,\text{m/sec}^2)(2.19\,\text{sec})^2$$

$$= 58.8\text{m}$$

Answer: 58.8 m

Alternatively, you can use the trajectory equation to calculate y when $x = 30$ m, $v_o = 40$ m/sec and $\theta = 70°$.

$$y = (\tan\theta)x - \frac{g}{2(v_o \cos\theta)^2} x^2$$

$$= (\tan 70)(30\text{m}) - \frac{9.81\frac{\text{m}}{\text{sec}^2}}{2(40\cos 70)^2}(30\text{m})^2$$

$$= 58.8\text{m}$$

2.5 PARTICLE'S HORIZONTAL PROJECTILE MOTION

In classical projectile motion, the object is thrown at an angle from the ground. Another kind of projectile motion is called Horizontal Projectile Motion, i.e., horizontally thrown projectile from a height. Let us assume a particle is thrown horizontally from point O with a velocity of v_o from a height of y, as shown in Figure 2.20.

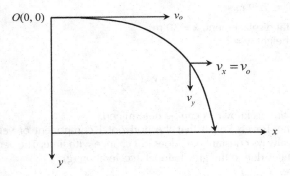

FIGURE 2.20   Travel Path of a Horizontal Projectile

As the particle is thrown horizontally at first, the vertical component of the initial velocity is zero ($v_{oy} = 0$). The horizontal component of the velocity is always the same, i.e., $v_x = v_o$. However, the vertical component of the velocity is zero at the beginning and increases due to the gravitational acceleration ($a_y = -g$) and reaches maximum just before touching the ground. Based on this discussion the following formulas can be derived:

Vertical acceleration, $a_y = -g$ (always acts downward and constant)
Horizontal acceleration, $a_x = 0$ (always zero for projectiles)
Horizontal velocity, $v_x = v_o$ (always constant for projectiles)
Vertical velocity, $v_y = v_{oy} + gt = 0 + gt = gt$ (decreases upward)
Horizontal displacement, $x = v_o t + \frac{1}{2}at^2 = v_o t$
Vertical displacement, $y = y_o + v_{oy}t + \frac{1}{2}gt^2 = \frac{1}{2}gt^2$

Rule: *Vertical velocity is zero at the beginning but develops due to gravity.*
Rule: *Gravitational acceleration always acts downward.*
Rule: *There is no horizontal acceleration for a projectile motion; thus, the horizontal velocity does not change.*
Rule: *At any point of the travel path, velocity has both x and y components.*

Example 2.24

A plane, 2 miles above the ground, is flying horizontally at a velocity of 500 m/sec. If it releases a bomb, calculate how much time the bomb will take to reach the ground, and how far the bomb will travel.

SOLUTION

Given,

Horizontal velocity, $v_x = v_o = 500$ m/sec

Vertical distance, $y = 2$ miles $\times \left(5,280 \dfrac{ft}{mile}\right) \times \left(0.3048 \dfrac{m}{ft}\right) = 3,219$ m

Time to reach the ground, $t = ?$

Known: $y = v_{oy}t + \dfrac{1}{2}gt^2$

Therefore, $3,219 \text{ m} = (0)(t) + \dfrac{1}{2}\left(9.81 \dfrac{m}{sec^2}\right)(t)^2$

$$3,219 \text{ m} = 4.905t^2$$

$$t = \sqrt{\dfrac{3,219}{4.905}} = 25.62 \text{ sec}$$

Horizontal displacement, $x = v_o t$

$$= \left(500\dfrac{m}{sec}\right)(25.62 \text{ sec})$$

$$= 12,810 \text{ m}$$

Answers:

25.63 sec
12,810 m

Example 2.25

A fighter plane while moving horizontally with a speed of 147 m/sec dropped a bomb. Ignoring the air resistance, calculate the velocity of the bomb after 5 sec of dropping.

SOLUTION

After 5 sec of dropping, the bomb will have both vertical and horizontal components of velocities. The horizontal component of the velocity does not change for a projectile.

Given,

Horizontal velocity, $v_x = v_o = 147 \text{m/sec}$
Time of travel, $t = 5$ sec
Combined velocity, $v = ?$

Known:

Horizontal velocity after 5 sec: $v_x = v_o = 147 \text{m/sec}$
Vertical velocity: $v_y = v_{oy} + gt = gt$

Vertical velocity after 5 sec: $v_y = gt = \left(9.81 \dfrac{\text{m}}{\text{sec}^2} \right)(5\,\text{sec}) = 49.05 \dfrac{\text{m}}{\text{sec}}$

The combined velocity, $v = \sqrt{v_x^2 + v_y^2} = \sqrt{\left(147 \dfrac{\text{m}}{\text{sec}} \right)^2 + \left(49.05 \dfrac{\text{m}}{\text{sec}} \right)^2} = 155 \dfrac{\text{m}}{\text{sec}}$

Answer: 155 m/sec

2.6 PLANAR CURVILINEAR MOTION

2.6.1 BACKGROUND

Curvilinear motion is defined as the motion when a particle moves along a curved path. This curved path can either be in two dimensions (on a plane), or in three dimensions. A curvilinear motion is more complex than rectilinear motion just discussed earlier in this chapter. This motion can be explained using three different coordinates systems namely Rectangular Components, Tangential and Normal Components, and Radial and Transverse Components.

2.6.2 RECTANGULAR COMPONENTS OF CURVILINEAR MOTION

The easiest way of representing curvilinear motion is the use of Cartesian coordinate or rectangular component systems. The motion (position, velocity, and acceleration) is expressed as a function of time in the Cartesian coordinate system. This system is good for motion in open areas such as an airplane. To obtain the velocity and acceleration of a particle undergoing curvilinear motion, we need to know the particle's position as a function of time. Let us assume a particle is moving along a non-straight path (curvilinear path), as shown in Figure 2.21, using the rectangular coordinates system. At a certain time, the position of the particle is $A(x, y)$. If the unit vectors along the x and y axes are $\vec{i}$ and $\vec{j}$, respectively, then the position of the particle can be defined by:

$$\vec{r} = x\vec{i} + y\vec{j}$$

 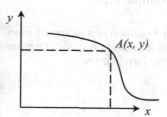

FIGURE 2.21 The Motion in a Curved Path using the Rectangular Components

The velocity can be expressed as:

$$\vec{v} = \frac{d\vec{r}}{dt} = \frac{d}{dt}\left(x\vec{i}\right) + \frac{d}{dt}\left(y\vec{j}\right) = v_x\vec{i} + v_y\vec{j}$$

The magnitude of the velocity can be determined as:

$$v = \sqrt{v_x^2 + v_y^2}$$

where:

$$v_x = \frac{dx}{dt} = \dot{x} = \text{velocity along the } x\text{-direction}$$

$$v_y = \frac{dy}{dt} = \dot{y} = \text{velocity along the } y\text{-direction}$$

Similarly, the acceleration can be written as:

$$\vec{a} = \frac{d\vec{v}}{dt} = a_x\vec{i} + a_y\vec{j}$$

The magnitude of the acceleration can be determined as:

$$a = \sqrt{a_x^2 + a_y^2}$$

where:

$$a_x = \frac{d^2x}{dt^2} = \ddot{x} = \text{acceleration along the } x\text{-direction}$$

$a_y = \dfrac{d^2y}{dt^2} = \ddot{y} =$ acceleration along the y-direction

Therefore, if the position of a particle as a function of time is known, to find the velocity and acceleration would be a fairly simple exercise. You simply take the first derivative to determine the velocity and the second derivative to determine the acceleration.

Example 2.26

The position of a jet can be represented as $(x, y) = (0.3t^3, 0.2t^3)$ where x and y are in m and t is in sec. Determine the velocity and the acceleration of the jet after traveling 5 sec.

SOLUTION

VELOCITY:

Velocity along the x-direction, $v_x = \dfrac{dx}{dt} = \dfrac{d}{dt}(0.3t^3) = 0.9t^2$

Velocity along the y-direction, $v_y = \dfrac{dy}{dt} = \dfrac{d}{dt}(0.2t^3) = 0.6t^2$
After $t = 5$ sec

Velocity along the x-direction, $v_x = 0.9t^2 = 0.9(5\,\text{sec})^2 = 22.5\,\text{m/sec}$

Velocity along the y-direction, $v_y = 0.6t^2 = 0.6(5\,\text{sec})^2 = 15\,\text{m/sec}$

The combined velocity, $v = \sqrt{v_x^2 + v_y^2} = \sqrt{\left(22.5\dfrac{\text{m}}{\text{sec}}\right)^2 + \left(15\dfrac{\text{m}}{\text{sec}}\right)^2} = 27\dfrac{\text{m}}{\text{sec}}$

ACCELERATION:

Acceleration along the x-direction, $a_x = \dfrac{d^2x}{dt^2} = \dfrac{d}{dt}(0.9t^2) = 1.8t$

Acceleration along the y-direction, $a_y = \dfrac{d^2y}{dt^2} = \dfrac{d}{dt}(0.6t^2) = 1.2t$
After $t = 5$ sec

Acceleration along the x-direction, $a_x = 1.8t = 1.8(5\,\text{sec}) = 9\,\dfrac{\text{m}}{\text{sec}^2}$

Acceleration along the y-direction, $a_y = 1.2t = 1.2(5\,\text{sec}) = 6\,\dfrac{\text{m}}{\text{sec}^2}$

The combined acceleration, $a = \sqrt{a_x^2 + a_y^2} = \sqrt{\left(9\,\dfrac{\text{m}}{\text{sec}^2}\right)^2 + \left(6\,\dfrac{\text{m}}{\text{sec}^2}\right)^2}$

$= 10.8\,\dfrac{\text{m}}{\text{sec}^2}$

Answers:

27 m/sec
10.8 m/sec²

Example 2.27

The end of a bar, P, slides along a fixed parabolic path with $y^2 = 20x$, where x and y are measured in cm, as shown in Figure 2.22. The y coordinate of P varies with time, t sec following $y = 4t^2 + 6t$ cm. Compute the velocity and acceleration of P when $t = 6$ sec.

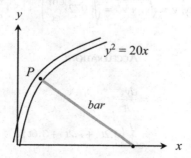

FIGURE 2.22   The Parabolic Path Traveled by the End of the Bar for Example 2.27

SOLUTION

VELOCITY:

Velocity along the x-direction, $v_x = \dfrac{dx}{dt}$

Velocity along the y-direction, $v_y = \dfrac{dy}{dt}$

y is expressed in terms of t. Now, let us express x in terms of t:

$$x = \frac{y^2}{20}$$

$$= \frac{\left(4t^2 + 6t\right)^2}{20}$$

$$= \frac{16t^4 + 48t^3 + 36t^2}{20}$$

$$= 0.8t^4 + 2.4t^3 + 1.8t^2$$

Therefore, $v_x = \dfrac{dx}{dt}$

$$= \frac{d}{dt}\left(0.8t^4 + 2.4t^3 + 1.8t^2\right)$$

$$= 3.2t^3 + 7.2t^2 + 3.6t$$

$$v_x\big|_{t=6\,\text{sec}} = 3.2\left(6\,\text{sec}\right)^3 + 7.2\left(6\,\text{sec}\right)^2 + 3.6\left(6\,\text{sec}\right) = 972\,\frac{\text{cm}}{\text{sec}}$$

$$v_y = \frac{dy}{dt} = \frac{d}{dt}\left(4t^2 + 6t\right) = 8t + 6$$

$$v_y\big|_{t=6\,\text{sec}} = 8\left(6\,\text{sec}\right) + 6 = 54 \text{ cm/sec}$$

The combined velocity, $v = \sqrt{v_x^2 + v_y^2} = \sqrt{\left(972\dfrac{\text{cm}}{\text{sec}}\right)^2 + \left(54\dfrac{\text{cm}}{\text{sec}}\right)^2} = 973.5 \dfrac{\text{cm}}{\text{sec}}$

ACCELERATION:

$$a_x = \frac{dv_x}{dt}$$

$$= \frac{d}{dt}\left(3.2t^3 + 7.2t^2 + 3.6t\right)$$

$$= 9.6t^2 + 14.4t + 3.6$$

$$a_x\big|_{t=6\,\text{sec}} = 9.6\left(6\,\text{sec}\right)^2 + 14.4\left(6\,\text{sec}\right) + 3.6 = 435.6 \frac{\text{cm}}{\text{sec}^2}$$

$$a_y = \frac{dv_y}{dt} = \frac{d}{dt}\left(8t + 6\right) = 8$$

$$a_y\big|_{t=6\,\text{sec}} = 8 \text{ cm/sec}^2$$

The combined acceleration, $a = \sqrt{a_x^2 + a_y^2} = \sqrt{\left(435.6\dfrac{\text{cm}}{\text{sec}^2}\right)^2 + \left(8\dfrac{\text{cm}}{\text{sec}^2}\right)^2}$
$= 435.7 \dfrac{\text{cm}}{\text{sec}^2}$

Answers:

973.5 cm/sec
435.7 cm/sec²

2.6.3 TANGENTIAL AND NORMAL COMPONENTS OF CURVILINEAR MOTION

The Cartesian coordinate system discussed above cannot represent some inherent properties of motion in a curvilinear path observed in our daily life such as driving a car, running, etc. Let us consider that curved path shown in Figure 2.23. If you want to drive very fast on this roadway, you may be thrown off the road due to developing centrifugal forces. The details of this centrifugal force are discussed in a future chapter. What is meant here is that the motion along this roadway cannot be fully described using the Cartesian coordinate. Now we will discuss about another way of representing curvilinear motion which is very applicable for roadways.

FIGURE 2.23 A Curved Path in CSU–Pueblo

Let a particle be moving in a curvilinear path, as shown in Figure 2.24, and its position at a certain time is A. If $\vec{u}_t$ and $\vec{u}_n$ are the unit vectors along the tangential and normal directions, respectively, then the velocity can be represented as:

$$\vec{v} = v_t \vec{u}_t$$

$$= \frac{ds}{dt} \vec{u}_t$$

$$= \dot{s} \vec{u}_t$$

Note that the velocity is always tangent to the path. It has no component along the normal direction.

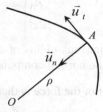

FIGURE 2.24 Concept of Tangential and Normal Components of Curvilinear Motion

Although velocity has no component along the normal direction, there is acceleration due to the change in direction of the velocity. The acceleration can be derived as:

$$\vec{a} = \frac{d\vec{v}}{dt}$$

$$= \frac{d}{dt}\left(v_t\vec{u}_t\right)$$

$$= \vec{u}_t\frac{d}{dt}\left(v_t\right) + v_t\frac{d}{dt}\left(\vec{u}_t\right)$$

$$= a_t\vec{u}_t + v_t\left(\frac{v_t}{\rho}\vec{u}_n\right)$$

Using calculus, it can be shown that $\frac{d}{dt}\left(\vec{u}_t\right) = \frac{v_t}{\rho}\vec{u}_n$. This derivation is skipped here to make it simple. Then, the combined acceleration can be written as:

$$\vec{a} = a_t\vec{u}_t + a_n\vec{u}_n$$

$$= a_t\vec{u}_t + \frac{v_t^2}{\rho}\vec{u}_n$$

where:
  ρ = Instantaneous radius of curvature

$$a_n = \frac{v_t^2}{\rho}$$

If the equation of the curve is known, the instantaneous radius of curvature can be determined as:

$$\rho = \frac{\left[1 + \left(\frac{dy}{dx}\right)^2\right]^{3/2}}{\left|\frac{d^2y}{dx^2}\right|}$$

For circular motion, the instantaneous radius of curvature is simply the radius of the circle, i.e., $\rho = R$. Then, the normal component of the acceleration will be $a_n = \frac{v_t^2}{R}$. The normal component of the force is then $F_n = ma_n$. This normal component of the force is also known as centripetal force. Therefore, centripetal force $F_n = ma_n = m\frac{v_t^2}{R} = \frac{mv_t^2}{R}$.

To clarify again, when a particle is in motion in a curved path, the velocity works along the tangent of the curve. The acceleration has two components: Tangential and normal. The tangential component of the acceleration is the rate of change in the magnitude of the velocity. Although there is no velocity along the normal direction, there is a normal component of acceleration. The normal component of acceleration is due to change in the direction of the velocity. If the velocity is constant, there is no tangential acceleration. However, there is the normal component of acceleration which always works toward the center of the curvature. This normal force, which is the product of the mass and the normal component of acceleration, is called the centripetal force. The details of the centripetal force and how it is handled in road design is discussed in Chapter 6.

Example 2.28

A car, as shown in Figure 2.25, is traveling along a path which can be expressed as $y = 0.1x^2$. After traveling 200 m, it attains a speed of 20 m/sec and an increase in speed of 2 m/sec^2 along the tangential direction. Calculate the magnitude of the acceleration of the car at this moment in time.

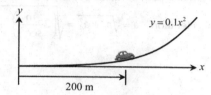

FIGURE 2.25 Position of a Car for Example 2.28

SOLUTION

The combined acceleration, $a = \sqrt{a_t^2 + a_n^2}$

Given the increase in speed is 2 m/sec^2. This is the tangential acceleration.

Therefore, tangential acceleration, $a_t = 2 \text{ m/sec}^2$

The normal acceleration can be written as $a_n = \dfrac{v_t^2}{\rho}$ where, ρ (instantaneous radius of curvature) is unknown.

$$\rho = \frac{\left[1+\left(\dfrac{dy}{dx}\right)^2\right]^{3/2}}{\left|\dfrac{d^2y}{dx^2}\right|}$$

Although there is no velocity along the normal direction, there is the normal component of acceleration. The normal component of acceleration occurs due to the change in direction of the velocity.

Given, $y = 0.1x^2$

Differentiating, $\dfrac{dy}{dx} = \dfrac{d}{dx}\left(0.1x^2\right) = 0.2x$

$$\frac{d^2y}{dx^2} = \frac{d}{dx}(0.2x) = 0.2$$

$$\frac{dy}{dx}\bigg|_{x=200m} = 0.2x = 0.2(200\,m) = 40$$

Therefore, $\rho = \dfrac{\left[1+\left(\dfrac{dy}{dx}\right)^2\right]^{3/2}}{\left|\dfrac{d^2y}{dx^2}\right|} = \dfrac{\left[1+(40)^2\right]^{3/2}}{|0.2|} = 320{,}300\ m$

Normal acceleration, $a_n = \dfrac{v_t^2}{\rho} = \dfrac{\left(20\,\dfrac{m}{sec}\right)^2}{320{,}300\ m} = 0.0012\ \dfrac{m}{sec^2}$

[Remember, for a circular path of radius R, $\rho = R$]

The combined acceleration, $a = \sqrt{a_t^2 + a_n^2} = \sqrt{\left(2\,\dfrac{m}{sec^2}\right)^2 + \left(0.0012\,\dfrac{m}{sec^2}\right)^2}$

$= 2.0\ \dfrac{m}{sec^2}$

Answer:

2.0 m/sec²

Example 2.29

The tangential acceleration of a car traveling in a curved path with radius of curvature 50 m can be represented as $\dot{v} = 0.03t^2$ where the distance is in m and t is in sec. Determine the velocity and the acceleration of the car after traveling 10 sec.

SOLUTION

Velocity after 10 sec:

Tangential acceleration $\dot{v} = a_t$

Tangential acceleration $a_t = \dfrac{dv}{dt}$
Then, $dv = a_t dt$
Let the velocities at time $= t_o$ and time $= t$ be v_o and v, respectively. Then,

on integrating, $\displaystyle\int_{v_o}^{v} dv = \int_{t_o}^{t} a_t dt$

$[v]_0^v = \displaystyle\int_{t_o}^{t} 0.03t^2 dt = 0.03\left[\frac{t^3}{3}\right]_{0\ sec}^{10\ sec}$ [given $t = 10$ sec]

Therefore, $v = 0.03\left[\dfrac{(10\sec)^3}{3} - \dfrac{(0\sec)^3}{3}\right] = 10 \text{ m/sec}$

Acceleration after 10 sec:

Tangential acceleration $a_t = \dot{v} = 0.03t^2 = 0.03(10\sec)^2 = 3\,\dfrac{\text{m}}{\sec^2}$

Normal acceleration $a_n = \dfrac{v_t^2}{\rho} = \dfrac{\left(10\dfrac{\text{m}}{\sec}\right)^2}{50 \text{ m}} = 2\,\dfrac{\text{m}}{\sec^2}$

The combined acceleration, $a = \sqrt{a_t^2 + a_n^2} = \sqrt{\left(3\dfrac{\text{m}}{\sec^2}\right)^2 + \left(2\dfrac{\text{m}}{\sec^2}\right)^2} = 3.6\,\dfrac{\text{m}}{\sec^2}$

Answers:

10 m/sec
3.6 m/sec²

2.6.4 RADIAL AND TRANSVERSE COMPONENTS OF CURVILINEAR MOTION

It is often convenient to express the planar (two–dimensional) motion of a particle using polar coordinates (r, θ), such that the velocity and acceleration of the particle in both the radial (r-direction) and the circumferential (θ-direction) directions can be determined. For such motion, a particle is considered to move along the radial r-direction for a given angle of θ. Suppose a particle is moving in a curvilinear path where r = radial position coordinate, θ = angle from the x-axis to $\vec{r}$, as shown in Figure 2.26. If $\vec{e}_r$ and $\vec{e}_\theta$ are the unit vectors with, and normal to the position vector, $\vec{r}$ respectively, then the position vector can be represented as:

$$\vec{r} = r\vec{e}_r$$

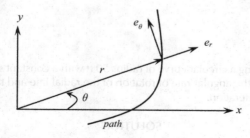

FIGURE 2.26 Demonstration of Radial and Transverse Components of a Curvilinear Motion

Note that the circumferential direction is perpendicular to the radial direction. The combined velocity can be represented as:

$$\vec{v} = v_r \vec{e}_r + v_\theta \vec{e}_\theta$$

$$= \dot{r}\vec{e}_r + r\dot{\theta}\vec{e}_\theta$$

$$v = \sqrt{(v_r)^2 + (v_\theta)^2}$$

The acceleration can be written as:

$$\vec{a} = (\ddot{r} - r\dot{\theta}^2)\vec{e}_r + (r\ddot{\theta} + 2\dot{r}\dot{\theta})\vec{e}_\theta$$

$$a = \sqrt{(a_r)^2 + (a_\theta)^2}$$

where
   r = radial position coordinate
   θ = angle from the x-axis to $\vec{r}$

$$v_r = \dot{r} = \frac{dr}{dt}$$

$$v_\theta = r\dot{\theta}$$

$$\dot{\theta} = \frac{d\theta}{dt}$$

$$\ddot{r} = \frac{d^2 r}{dt^2}$$

$$a_r = \ddot{r} - r\dot{\theta}^2$$

$$a_\theta = r\ddot{\theta} + 2\dot{r}\dot{\theta}$$

Example 2.30

A car travels along a circular curve of radius 50 ft with a constant speed of 10 ft/sec. Determine the angular rate of rotation of the radial line and the magnitude of the car's acceleration.

SOLUTION

Speed, v = 10 ft/sec
Radius, r = 50 ft

$$\dot{r} = \frac{dr}{dt} = 0$$

$$\ddot{r} = \frac{d^2r}{dt^2} = 0$$

$$v_r = \dot{r} = 0$$

$$v_\theta = r\dot{\theta} = 50(\dot{\theta})$$

Therefore, $v = \sqrt{(v_r)^2 + (v_\theta)^2} = \sqrt{(0)^2 + (50\dot{\theta})^2} = 10\ \dfrac{ft}{sec}$

Solving $\dot{\theta} = 0.2\ \dfrac{rad}{sec}$

$$\ddot{\theta} = \frac{d}{dt}\dot{\theta} = 0$$

$$a_r = \ddot{r} - r\dot{\theta}^2 = 0 - 50(0.2)^2 = -2\ \frac{ft}{sec^2}$$

$$a_\theta = r\ddot{\theta} + 2\dot{r}\dot{\theta} = 50(0) + 2(0)(0.2) = 0$$

$$a = \sqrt{(a_r)^2 + (a_\theta)^2} = \sqrt{\left(-2\ \frac{ft}{sec^2}\right)^2 + \left(0\ \frac{ft}{sec^2}\right)^2} = 2\ \frac{ft}{sec^2}$$

Answers:

10 ft/sec
2.0 ft/sec²

Example 2.31

A car, as shown in Figure 2.27, is traveling along a path which can be expressed as $r = 10(\theta^2)$, where θ is in radians. If it runs at a speed of 30 ft/sec, determine the radial and the transverse components of its velocity when it travels $\theta = \pi/2$ rad.

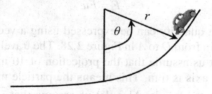

FIGURE 2.27 Position of a Car for Example 2.31

SOLUTION

Given, $r = 10(\theta^2)$

Speed, $v = 30$ ft/sec
Travel path, $\theta = \pi/2$ rad

Then, $r\big|_{\theta=\frac{\pi}{2}} = 10(\theta^2) = 10\left(\frac{\pi}{2}\right)^2 = 24.7$ ft

$$v_r = \dot{r} = 10(2\theta)\dot{\theta}$$

Then, velocity along the radius, $v_r\big|_{\theta=\frac{\pi}{2}} = \dot{r}\big|_{\theta=\frac{\pi}{2}} = (20\theta)\dot{\theta} = 20\left(\frac{\pi}{2}\right)\dot{\theta} = 31.4\dot{\theta}$

Velocity along the θ, $v_\theta\big|_{\theta=\frac{\pi}{2}} = r\dot{\theta} = 10\left(\frac{\pi}{2}\right)^2(\dot{\theta}) = 24.7\dot{\theta}$

Now, $v^2 = (v_r)^2 + (v_\theta)^2$

$$(30)^2 = (31.4\dot{\theta})^2 + (24.7\dot{\theta})^2$$

$$\dot{\theta} = 0.75\,\frac{rad}{sec}$$

Therefore,

Velocity along the radius, $v_r = 31.4\dot{\theta} = 31.4\left(0.75\,\frac{rad}{sec}\right) = 23.6\,\frac{ft}{sec}$

Velocity along the θ, $v_\theta = 24.7\dot{\theta} = 24.7\left(0.75\,\frac{rad}{sec}\right) = 18.5\,\frac{ft}{sec}$

Answers:

23.6 ft/sec
18.5 ft/sec

Knowledge on vectors is required to move forward. A vector, $\vec{F}$ can be determined as the product of the scalar quantity (magnitude of the force, F) times the unit vector, $\vec{u}$ along the direction as follows:

$$\vec{F} = F\vec{u}$$

Now let us see how a quantity can be expressed using a vector. Let us consider that a car traveled 10 m from O to A in Figure 2.28. The travel path lies in between the x and y axes. Let us assume that the projection of 10 m along the x-axis is 8 m, and along the y axis is 6 m. This means the particle moved 8 m along the x-axis and 6 m along the y axis. Also, let us assume that $\vec{i}$ represents the unit vector along the x-axis, meaning a magnitude of one (1.0) along the x-axis. More

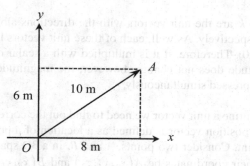

FIGURE 2.28   Representation of a Vector

clearly, $\vec{i}$ represents a quantity with a magnitude of one (1.0) and acts along the x-axis. Similarly, let us assume that $\vec{j}$ represents the unit vector along the y axis. Unit vector ($\vec{u}$) means a vector that shows the direction with a magnitude of 1.0. Therefore, multiplying a unit vector with a scalar force does not change the magnitude of the force; it just adds the direction with it.

Then, the displacement can be presented as $8\vec{i} + 6\vec{j}$ m. Thus, any vector quantity can be represented as:

$$\vec{F} = F_x\vec{i} + F_y\vec{j}$$

The resultant direction (θ_x) with respect to the x-axis is

$$\theta_x = \tan^{-1}\left(\frac{F_y}{F_x}\right)$$

Now let us discuss the scenario for a 3D system where there are three dimensions of the system (x, y, and z).

In a 3D system, $\vec{F} = F_x\vec{i} + F_y\vec{j} + F_z\vec{k}$

The magnitude of $\vec{F}$ can be determined as: $F = \sqrt{\left(F_x\right)^2 + \left(F_y\right)^2 + \left(F_z\right)^2}$

where:

$F_x = F\cos\theta_x$ = the x component of the quantity F
$F_y = F\cos\theta_y$ = the y component of the quantity F
$F_z = F\cos\theta_z$ = the z component of the quantity F

$\cos\theta_x = \dfrac{F_x}{F}$ = the angle between the x-axis and the direction of the quantity F

$\cos\theta_y = \dfrac{F_y}{F}$ = the angle between the y axis and the direction of the quantity F

$\cos\theta_z = \dfrac{F_z}{F}$ = the angle between the z axis and the direction of the quantity F

$\vec{i}, \vec{j}$ and $\vec{k}$ are the unit vectors with the directions along the x, y and z axes, respectively. As well, each of these unit vectors has a magnitude of one (1.0). Therefore, if it is multiplied with a scalar quantity, then the magnitude does not change, however, the magnitude and direction can be expressed simultaneously.

To determine a unit vector we need to find out the corresponding position vector first. A position vector is defined as a location of a point in space relative to another point. Consider two points, A and B, in a 3D space, as shown in Figure 2.29. Let their coordinates be $A(x_A, y_A, z_A)$ and $B(x_B, y_B, z_B)$, respectively. The positive vector ($\vec{P}_{AB}$) directed from A to B can be determined as follows:

$$\vec{P}_{AB} = (x_B - x_A)\vec{i} + (y_B - y_A)\vec{j} + (z_B - z_A)\vec{k}$$

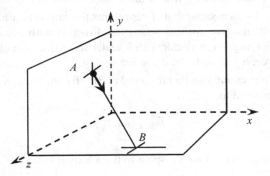

FIGURE 2.29   Vector Method of Expressing the Positions

Note that A is the starting point and B is the ending point. Always subtract the end coordinates from the start coordinates to determine the position vector. Then the position vector is divided by its magnitude to determine the unit vector ($\vec{u}_{AB}$) as follows:

$$\vec{u}_{AB} = \frac{\vec{P}_{AB}}{|P_{AB}|}$$

$$\vec{u}_{AB} = \frac{(x_B - x_A)\vec{i} + (y_B - y_A)\vec{j} + (z_B - z_A)\vec{k}}{\sqrt{(x_B - x_A)^2 + (y_B - y_A)^2 + (z_B - z_A)^2}}$$

If there are several vectors, $\vec{F}_1 = F_{1x}\vec{i} + F_{1y}\vec{j} + F_{1z}\vec{k}$, $\vec{F}_2 = F_{2x}\vec{i} + F_{2y}\vec{j} + F_{2z}\vec{k}$, etc., acting simultaneously at a point then, their resultant ($\vec{R}$) can be determined as summing up the corresponding components as follows:

$$\vec{R} = (F_{1x} + F_{2x} + ...)\vec{i} + (F_{1y} + F_{2y} + ...)\vec{j} + (F_{1z} + F_{2z} + ...)\vec{k}$$

The magnitude of the resultant (R) can be determined as:

$$R = \sqrt{\left(F_{1x} + F_{2x} + ...\right)^2 + \left(F_{1y} + F_{2y} + ...\right)^2 + \left(F_{1z} + F_{2z} + ...\right)^2}$$

2.7 RELATIVE MOTION

The relative motion of a moving object is determined with respect to another moving object. Thus, the motion is not determined with reference to the earth but as the motion of the object with regard to the other moving object as if it were static. For example, a person sitting in a bus is at zero velocity relative to the bus while moving at the same velocity as an airplane that is taking off with respect to the ground. The calculation of relative motion is carried out in terms of relative velocity, relative speed, or relative acceleration (the change in velocity divided by the change in time).

Consider particles A and B are moving independently along two paths with positions of $\vec{r}_A$ and $\vec{r}_B$ respectively, as shown in Figure 2.30. The position of each particle is measured from the common origin O of the fixed XY reference axis. The origin of the second frame of reference xy is attached to particle B and moves with it with a position of $\vec{r}_B$. The axes of this frame are only permitted to translate relative to the fixed frame, XY. The relative position of A with respect to B is designated by a relative vector, $\vec{r}_{A/B}$.

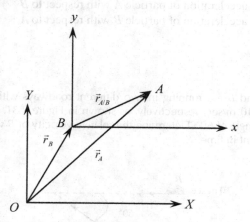

FIGURE 2.30 Demonstration of Relative Motion of a Particle A with Respect to Particle B

Then, the position vector of A can be related as follows:

$$\vec{r}_A = \vec{r}_B + \vec{r}_{A/B}$$

The relative velocity and relative acceleration of particle A with respect to B can be expressed as follows:

$$\vec{v}_A = \vec{v}_B + \vec{v}_{A/B}$$

$$\vec{a}_A = \vec{a}_B + \vec{a}_{A/B}$$

The relative velocity and relative acceleration of particle B with respect to A can be expressed as follows:

$$\vec{v}_B = \vec{v}_A + \vec{v}_{B/A}$$

$$\vec{a}_B = \vec{a}_A + \vec{a}_{B/A}$$

where:
$\vec{r}_A$ = position vector of particle A
$\vec{r}_B$ = position vector of particle B
$\vec{r}_{A/B}$ = relative position vector of particle A with respect to B
$\vec{r}_{B/A}$ = relative position vector of particle B with respect to A
$\vec{v}_A$ = velocity of particle A
$\vec{v}_B$ = velocity of particle B
$\vec{v}_{A/B}$ = relative velocity of particle A with respect to B
$\vec{v}_{B/A}$ = relative velocity of particle B with respect to A
$\vec{a}_A$ = acceleration of particle A
$\vec{a}_B$ = acceleration of particle B
$\vec{a}_{A/B}$ = relative acceleration of particle A with respect to B
$\vec{a}_{B/A}$ = relative acceleration of particle B with respect to A

Example 2.32

Two cars, A and B are running in two different roadways with velocities of 14 m/sec and 10 m/sec, respectively, as shown in Figure 2.31. The roads are aligned at an angle of 60°. Determine the relative velocity of B with respect to A at this moment in time.

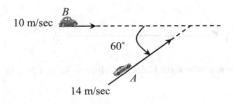

10 m/sec

60°

14 m/sec

FIGURE 2.31   Positions of Two Cars for Example 2.32

SOLUTION

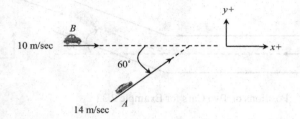

FIGURE 2.32 Redrawing of the Positions of the Two Cars

Let $\vec{i}$ and $\vec{j}$ be the unit vectors along the x and y axes, as shown in Figure 2.32.

Velocity of A in vector: $\vec{v}_A = 14\cos60\,\vec{i} + 14\sin60\,\vec{j}\ \dfrac{m}{sec}$

Velocity of B in vector: $\vec{v}_B = 10\vec{i}\ \dfrac{m}{sec}$

Known equation: $\vec{v}_B = \vec{v}_A + \vec{v}_{B/A}$
Therefore, $\vec{v}_{B/A} = \vec{v}_B - \vec{v}_A$

$$= 10\vec{i} - \left(14\cos60\,\vec{i} + 14\sin60\,\vec{j}\right)$$

$$= 3\vec{i} - 12.12\vec{j}$$

Magnitude of $v_{B/A} = \sqrt{\left(3\,\dfrac{m}{sec}\right)^2 + \left(-12.12\,\dfrac{m}{sec}\right)^2} = 12.5\,\dfrac{m}{sec}$

Direction of $\vec{v}_{B/A}$, $\theta = \tan^{-1}\left(\dfrac{12.12}{3}\right) = 76°$

The coefficients of $\vec{i}$ and $\vec{j}$ in the equation $\vec{v}_{B/A} = 3\vec{i} - 12.12\vec{j}$ are positive ($\rightarrow$) and negative ($\downarrow$), respectively. This means the angle (θ) is located in 4th quadrant. The order of the quadrant is counted counter-clockwise. Therefore, the angle (θ) is clockwise with reference to the positive x-axis. In other words, it can be said that the angle (θ) is $360° - 76° = 284°$ counter-clockwise with reference to the positive x-axis.

Example 2.33

At a given instant, as shown in Figure 2.33, the car at A is traveling at 14 m/sec with acceleration of 2 m/sec² and the car at B is traveling at 10 m/sec with acceleration of 3 m/sec². Determine the relative velocity and the relative acceleration of A with respect to B at this moment in time, and their directions.

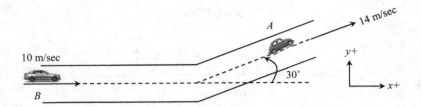

FIGURE 2.33   Positions of Two Cars for Example 2.33

SOLUTION

Assume unit vectors $\vec{i}$ and $\vec{j}$ along the horizontal and vertical axes, respectively.

RELATIVE VELOCITY:

Velocity of A in vector: $\vec{v}_A = 14\cos 30\vec{i} + 14\sin 30\vec{j} = 12.12\vec{i} + 7\vec{j}$

Velocity of B in vector: $\vec{v}_B = 10\vec{i}\ \dfrac{m}{sec}$

Known equation: $\vec{v}_A = \vec{v}_B + \vec{v}_{A/B}$

Therefore, $\vec{v}_{A/B} = \vec{v}_A - \vec{v}_B$

$$= 12.12\vec{i} + 7\vec{j} - 10\vec{i}$$

$$= 2.12\vec{i} + 7\vec{j}$$

$$v_{A/B} = \sqrt{\left(2.12\frac{m}{sec}\right)^2 + \left(7\frac{m}{sec}\right)^2} = 7.31\frac{m}{sec}$$

$$\theta = \tan^{-1}\left(\frac{7}{2.12}\right) = 73.2^o$$

The coefficients of $\vec{i}$ and $\vec{j}$ in the equation $\vec{v}_{A/B} = 2.12\vec{i} + 7\vec{j}$ are positive ($\rightarrow$, ↑). This means the angle (θ) is located in the 1st quadrant. Therefore, the angle (θ), 73.2° is counter-clockwise with reference to the positive x-axis.

RELATIVE ACCELERATION:

Acceleration of A in vector: $\vec{a}_A = 2\cos 30\vec{i} + 2\sin 30\vec{j}\ \dfrac{m}{sec^2}$

$$= 1.73\vec{i} + \vec{j}\ \frac{m}{sec^2}$$

Acceleration of B in vector: $\vec{a}_B = 3\vec{i}\ \dfrac{m}{sec^2}$

Known equation: $\vec{a}_A = \vec{a}_B + \vec{a}_{A/B}$

Therefore, $\vec{a}_{A/B} = \vec{a}_A - \vec{a}_B$

$$= 1.73\vec{i} + \vec{j} - 3\vec{i}$$

$$= -1.27\vec{i} + \vec{j}$$

$$a_{A/B} = \sqrt{\left(-1.27\frac{m}{sec^2}\right)^2 + \left(1\frac{m}{sec^2}\right)^2} = 1.62\frac{m}{sec^2}$$

$$\theta = \tan^{-1}\left(\frac{1}{1.27}\right) = 38°$$

The coefficients of $\vec{i}$ and $\vec{j}$ in the equation $\vec{a}_{A/B} = -1.27\vec{i} + \vec{j}$ are negative (←) and positive (↑), respectively. This means the angle (θ) is located in the 2nd quadrant. Therefore, the angle (θ), 38° is clockwise with reference to the negative x-axis. In other words, 180° − 38° = 142° counter-clockwise with reference to the positive x-axis.

Answers:

   7.31 m/sec, 73.15° counter-clockwise with respect to the positive x-axis
   1.62 m/sec², 38° clockwise with respect to the negative x-axis

The above two examples discuss the relative motion where both particles are in motion along straight paths although in different directions. Now, we will discuss if the motion is in a curved path.

Example 2.34

At the instant shown in Figure 2.34 the car at A is traveling at 14 m/sec with acceleration of 2 m/sec² around the curve with the radius of curvature of 50 m and the car at B is traveling at 10 m/sec with acceleration of 3 m/sec² along the straightway. Determine the relative velocity and relative acceleration of A with respect to B at this moment in time.

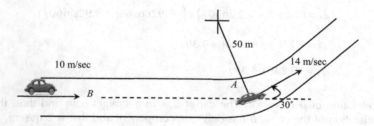

FIGURE 2.34 Positions of Two Cars for Example 2.34

SOLUTION

Assume unit vectors $\vec{i}$ and $\vec{j}$ along the horizontal and vertical axes, respectively.

Velocity of A in vector: $\vec{v}_A = 14\cos 30\vec{i} + 14\sin 30\vec{j} = 12.12\vec{i} + 7\vec{j}$

Velocity of B in vector: $\vec{v}_B = 10\vec{i} \dfrac{m}{\sec}$

Known equation: $\vec{v}_A = \vec{v}_B + \vec{v}_{A/B}$

Therefore, $\vec{v}_{A/B} = \vec{v}_A - \vec{v}_B = 12.12\vec{i} + 7\vec{j} - 10\vec{i} = 2.12\vec{i} + 7\vec{j}$

$$v_{A/B} = \sqrt{\left(2.12\frac{m}{\sec}\right)^2 + \left(7.0\frac{m}{\sec}\right)^2} = 7.31\frac{m}{\sec}$$

$$\theta = \tan^{-1}\left(\frac{7.0\dfrac{m}{\sec}}{2.12\dfrac{m}{\sec}}\right) = 73.2^{\circ}$$

The coefficients of $\vec{i}$ and $\vec{j}$ in the equation $\vec{v}_{A/B} = 2.12\vec{i} + 7\vec{j}$ are positive ($\rightarrow$, $\uparrow$). This means the angle (θ) is located in the 1st quadrant. Therefore, the angle (θ), 73.2° is counter-clockwise with reference to the positive x-axis.

Acceleration of car at A in vector:

As the car at A is in a curve, it has both normal and tangential components of acceleration. The acceleration of the car at A is given as 2 m/sec². This is the tangential component of the acceleration. The normal components of the acceleration can be calculated as:

$$\left(a_A\right)_n = \frac{v_A^2}{\rho} = \frac{\left(14 \text{ m/sec}\right)^2}{50\text{m}} = 3.92\frac{m}{\sec^2}$$

$$\left(a_A\right)_t = 2\frac{m}{\sec^2}$$

From Figure 2.35:

$$\vec{a}_A = \left(2\cos 30\vec{i} + 2\sin 30\vec{j}\right) + \left(-3.92\cos 60\vec{i} + 3.92\sin 60\vec{j}\right)$$

$$= 1.73\vec{i} + \vec{j} - 1.96\vec{i} + 3.39\vec{j}$$

$$= -0.23\vec{i} + 4.39\vec{j}$$

Acceleration of B in vector: The car at B is in a straight path and thus, the acceleration of the car at B has only one component and that is 2 m/sec². In vector form:

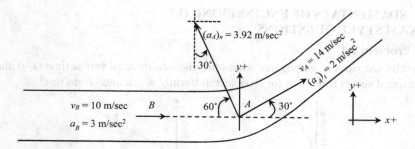

FIGURE 2.35 Orientation of Velocities and Accelerations at A and B

$$\vec{a}_B = 3\vec{i}\ \text{m/sec}^2$$

Known equation: $\vec{a}_A = \vec{a}_B + \vec{a}_{A/B}$

Therefore, $\vec{a}_{A/B} = \vec{a}_A - \vec{a}_B$

$$= -0.23\vec{i} + 4.39\vec{j} - 3\vec{i}$$

$$= -3.23\vec{i} + 4.39\vec{j}$$

$$a_{A/B} = \sqrt{\left(-3.23\frac{\text{m}}{\text{sec}^2}\right)^2 + \left(4.39\frac{\text{m}}{\text{sec}^2}\right)^2} = 5.45\ \frac{\text{m}}{\text{sec}^2}$$

$$\theta = \tan^{-1}\left(\frac{4.39\ \dfrac{\text{m}}{\text{sec}^2}}{3.23\ \dfrac{\text{m}}{\text{sec}^2}}\right) = 53.7^\circ$$

The coefficients of $\vec{i}$ and $\vec{j}$ in the equation $\vec{a}_{A/B} = -3.23\vec{i} + 4.39\vec{j}$ are negative ($\leftarrow$) and positive ($\uparrow$), respectively. This means the angle (θ) is located in the 2nd quadrant. Therefore, the angle (θ), 53.7° is clockwise with reference to the negative x-axis. In other words, 180° − 53.7° = 126.3° counter-clockwise with reference to the positive x-axis.

Answers:

7.31 m/sec, 73.15° counter-clockwise with respect to the positive x-axis
5.45 m/sec², 53.7° clockwise with respect to the negative x-axis

In this example, only one particle is in motion in a curved path; the other particle is in motion along a straight path. If both particles are in motion in curved paths, then accelerations of both particles are to be determined considering the normal and tangential components.

FUNDAMENTALS OF ENGINEERING (FE) EXAM STYLE QUESTIONS

FE Problem 2.1

Identify the pairs of graphs that shows the distance traveled versus time (s–t) and the speed versus time (v–t) for an object uniformly accelerated from rest?

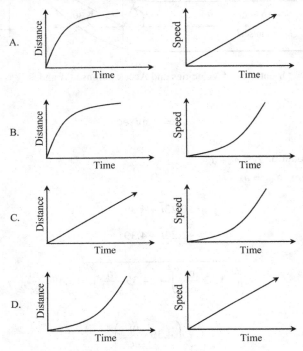

FE Problem 2.2

At time $t = 0$, car P traveling with speed v_o passes car Q that just started to move. Now both cars are traveling on two parallel lanes along the same straight road. The speed versus time graph for both cars are shown in Figure 2.36. Identify the correct statement at time $t = 20$ sec.

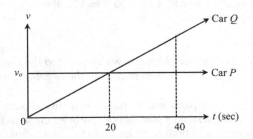

FIGURE 2.36 The v–t Curve of Two Cars for FE Problem 2.2

A. Car Q is passing car P
B. Car Q is behind car P
C. Car Q is in front of car P
D. Car P is accelerating faster than car Q

FE Problem 2.3

A car driver makes an emergency stop by slamming on the brakes and skidding to a stop. How far did the car skid if it had been traveling twice as fast?

A. 4 times as far
B. the same distance
C. 2 times as far
D. the mass of the car must be known to calculate

FE Problem 2.4

The velocity (v) vs. time (t) graph for the motion of a car on a straight track is shown in Figure 2.37. Assume that the car starts from the origin where $x = 0$. Determine the time (sec) that the car will be at the greatest distance from the origin?

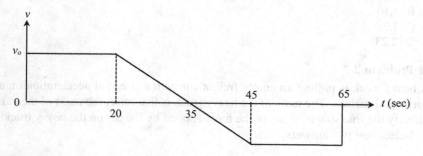

FIGURE 2.37   The v–t Curve for FE Problem 2.4

A. 20
B. 35
C. 45
D. 65

FE Problem 2.5

A particle is moving with a velocity of 2 m/sec on a straight line with a deceleration of 0.1 m/sec². The time (sec) required to travel a distance of 60 m is most nearly:

A. 40
B. 30
C. 20
D. 15
E. 10

FE Problem 2.6

A 0.20 kg object is moving along a straight path. The net force that acts on the object varies with the object's displacement, as shown in Figure 2.38. The object starts from rest at displacement $x = 0$ and time $t = 0$ and is displaced a distance of 20 m. The time (sec) taken for the object to be displaced the first 12 m is most nearly:

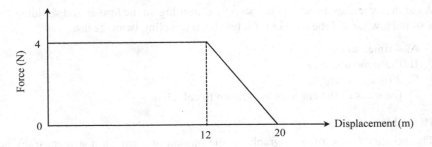

FIGURE 2.38   The Force–Displacement Curve for FE Problem 2.6

A. 0.12
B. 1.10
C. 1.98
D. 2.23

FE Problem 2.7

A heavy truck is pulling an empty freight car with a constant acceleration on a horizontal surface. The mass of the heavy truck is five times the mass of the car. Identify the true statement about the force applied by the car on the heavy truck.
    [select best two answers]

A. Five times greater than the force applied by the heavy truck on the car
B. Five times less than the force applied by the heavy truck on the car
C. The force applied by the car is zero since they both move with constant accelerations
D. equal to the force of the heavy truck on the car
E. opposite in direction to the force applied by the heavy truck on the car

FE Problem 2.8

The motion of a particle is described by the relation $s = t^2 - 10t + 30$, where s is in m and t is in sec. The total distance traveled by the particle from $t = 0$ to $t = 10$ sec would be most nearly:

A. zero
B. 30 m
C. 50 m
D. 60 m
E. none of these options

FE Problem 2.9

The displacement of a particle that moves along a straight line is given by $s = 4t^3 + 3t^2 - 10$ where s is in m and t is in sec. The time (sec) taken by the particle to acquire a velocity of 18 m/sec from rest is most nearly:

    A. 0.5
    B. 1.0
    C. 1.5
    D. −1.5

FE Problem 2.10

The time (sec) required to stop a car moving with a velocity 20 m/sec within a distance of 40 m is most nearly:

    A. 2
    B. 3
    C. 4
    D. 5
    E. 6

FE Problem 2.11

A projectile is fired with an initial velocity of v_o and at an angle θ_o with the horizontal and follows the projectile trajectory commonly known. Consider, the vertical upward direction is positive. Identify the pairs of graphs from the list below that best represents the vertical components of the velocity and acceleration, v_y and a_y, respectively, of the projectile as functions of time, t.

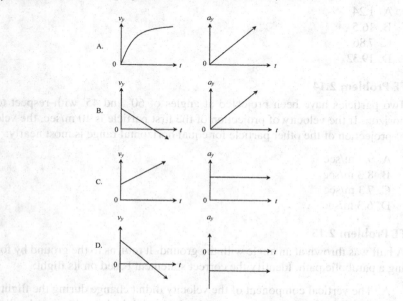

FE Problem 2.12

A ball is thrown from location O and follows the parabolic path, as shown in Figure 2.39. Consider that point Q is the highest point on the parabolic path, and points P and R are at the same level above the ground. If the combined velocity (magnitude only) at P, Q, and R are v_P, v_Q, and v_R, respectively, how do the speeds of the ball at these three points compare? Consider that the friction of air is negligible.

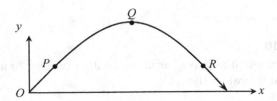

FIGURE 2.39   The Flow Path of a Projectile for FE Problem 2.12

A. $v_P < v_Q < v_R$
B. $v_P > v_Q > v_R$
C. v_P and v_R are equal and they both are greater than v_Q
D. v_P and v_R are equal and they both are smaller than v_Q

FE Problem 2.13

A projectile is launched with a velocity of 100.3 m/sec at an angle of 60° with respect to the surface. The time (sec) required by the projectile to reach a height of 100 m from the ground while returning is most nearly:

A. 1.24
B. 16.5
C. 7.86
D. 19.32

FE Problem 2.14

Two particles have been projected at angles of 60° and 45° with respect to the horizon. If the velocity of projection of the first particle is 10 m/sec, the velocity of projection of the other particle for equal horizontal range is most nearly:

A. 9.3 m/sec
B. 8.3 m/sec
C. 7.3 m/sec
D. 6.3 m/sec

FE Problem 2.15

A ball was thrown at an angle with the ground. It returns to the ground by following a parabolic path. Identify the correct statement based on its flight:

A. The vertical component of the velocity didn't change during the flight
B. The horizontal component of the velocity didn't change during the flight

C. The speed of the ball was always the same

D. The kinetic energy of the ball was always the same during the flight

FE Problem 2.16

A projectile was launched with a velocity of 100.3 m/sec at an angle of 60° with the ground. The velocity (m/sec) attained by the projectile when it is moving at a height of 100 m is most nearly:

A. 70

B. 75

C. 80

D. 85

E. 90

FE Problem 2.17

A car is moving at a constant velocity on a gravel road. A small rock gets stucks in a groove of one of the tires. What would be the direction of the initial acceleration of the rock at the moment it gets attached to the tire?

A. vertically upward

B. horizontally forward

C. horizontally backward

D. inclined backward

PRACTICE PROBLEMS

Section I: Analytical Analysis of Particle's Rectilinear Motion

Problem 2.1

A particle's velocity is represented by $v(t) = 22 + 2t$ m/sec, t is in sec. Calculate the particle's travel distance after 10 sec.

Problem 2.2

A man can swim directly across a river of 50 m width in 4 min if there is no current. He takes 5 min to cross directly when there is current. Calculate the velocity of the current.

Problem 2.3

A 10-kg block begins to slide down in a 25° inclined plane. Starting from rest, the block attains a velocity of 5 m/sec after 5 sec. Calculate how far the block travels in that 5 sec.

Problem 2.4

A car starts from rest from Pueblo City Office and accelerates at 0.1 m/sec² for 5 min. Afterwards it travels with a constant velocity for 40 min. It then decelerates at 0.1 m/sec² until it is brought to rest in Colorado Springs City Office. Determine the distance between Pueblo City Office and Colorado Springs City Office.

Section II: Graphical Analysis of Particle's Rectilinear Motion

Problem 2.5

The position–time distribution of a hypothetical speed train is shown in Figure 2.40.

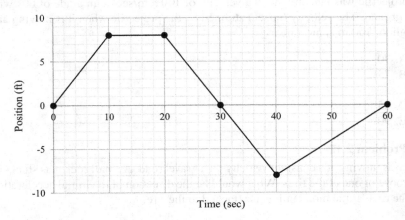

FIGURE 2.40 The s–t Curve for Problem 2.5

Calculate the velocity at 15 sec.

Calculate the velocity at 50 sec.

Calculate the distance traveled in 30 sec.

Calculate the displacement in 30 sec.

Calculate the acceleration at 50 sec.

Problem 2.6

The velocity–time distribution of a hypothetical speed train is shown in Figure 2.41.

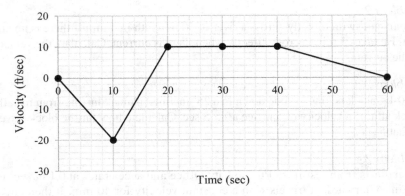

FIGURE 2.41 The v–t Curve for Problem 2.6

Calculate the acceleration at 15 sec.
Calculate the acceleration at 30 sec.
Calculate the distance traveled in 60 sec.
Calculate the displacement in 60 sec.

Problem 2.7

The acceleration–time distribution of a hypothetical speed train is shown in Figure 2.42.

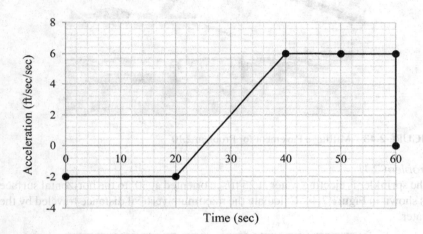

FIGURE 2.42 The a–t Curve for Problem 2.7

Calculate the velocity at 15 sec.
Calculate the velocity at 30 sec.
Calculate the displacement in 60 sec.
Calculate the distance traveled in 60 sec.

Section III: Projectile Motions

Problem 2.8

A ball is thrown vertically in the upward direction from the top of a tower with a velocity of 14.5 m/sec. It reaches the bottom of the tower after 5 sec. Calculate the height of the tower.

Problem 2.9

A rock is thrown vertically upward with a velocity of 4.9 m/sec. Determine the following:

(a) The velocity when it reaches the maximum height
(b) The maximum height it reaches vertically
(c) The total flight time

Problem 2.10
A milling excavator throws the milling at 15 ft/sec at an angle of 20° with the
surface, as shown in Figure 2.43. Calculate the maximum horizontal distance the
milling will travel if the vertical fall of the millings is 12 ft.

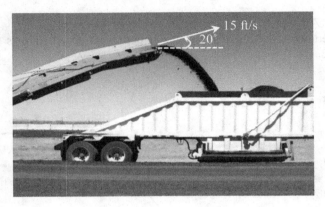

FIGURE 2.43   Milling a Pavement for Problem 2.10

Problem 2.11
The sprinkler is ejecting water at 25 ft/sec oriented at 30° to the horizontal surface
as shown in Figure 2.44. Calculate the maximum vertical distance traveled by the
water.

FIGURE 2.44   Sprinkling a Playground at Farmingdale State College for Problem 2.11

Problem 2.12
A particle is projected with a velocity of 20 m/sec with an angle of 45° to the
horizon. Determine the following:

(a) the maximum height
(b) the time of flight
(c) horizontal range

Problem 2.13
A boy has thrown a stone at an initial velocity of 25 ft/sec launched an angle of
15° down with the horizon from the top of a building, as shown in Figure 2.45.
Calculate the maximum horizontal distance traveled by the stone.

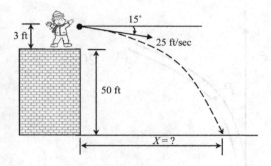

FIGURE 2.45   Throwing a Stone for Problem 2.13

Problem 2.14
A fighter plane while moving horizontally at a height of 1,000 m with a speed of 200 m/sec dropped a bomb. Ignoring the air resistance, calculate the velocity after 10 sec of dropping.

Problem 2.15
A fighter plane while moving horizontally at a height of 490 m and a speed of 147 m/sec dropped a bomb. Ignoring the air resistance, calculate when the bomb will fall on the ground.

Problem 2.16
A fighter plane while moving horizontally at a height of 490 m with a speed of 147 m/sec dropped a bomb. Ignoring the air resistance, calculate where the bomb will fall on the ground from the dropping point.

Problem 2.17
A balloon is moving horizontally with a uniform velocity of 196 m/sec and a stone is let fall down from the balloon which reaches the ground in 5 sec. Calculate the velocity of the stone when it is at the moment of touching the ground.

Section IV: Curvilinear Motion
Problem 2.18
The end of a bar, P slides along a fixed parabolic path with $y^2 = 20x$, where x and y are measured in cm as shown in Figure 2.46. The y coordinate of P varies with time, t sec following $y = 4t^2 + 6t$ cm. Compute the velocity and acceleration of P when $t = 6$ sec.

Problem 2.19
The tangential acceleration of a car travelling in a curved path of radius of curvature of 50 m can be represented as $\dot{v} = 0.03t^2$ where the distance is in m and t is in sec. Determine the velocity and the acceleration of it after travelling 10 sec.

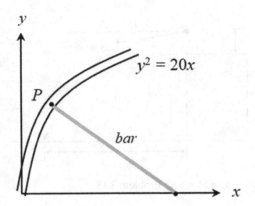

FIGURE 2.46   The parabolic path travelled by the end of the bar for Problem 2.18

Problem 2.20

A car as shown in Figure 2.47, is travelling along a path which can be expressed as $r = 10\left(\theta^2\right)$, where θ is in radians. If it runs at a speed of 30 ft/sec, determine the radial and the transverse components of its velocity when it travels $\theta = \pi/2$.

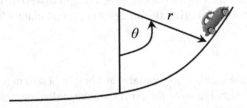

FIGURE 2.47   Position of a car for Problem 2.20

Section V: Relative Motion

Problem 2.21

At the instant shown in Figure 2.48, the car at A is traveling at 20 ft/sec with acceleration of 4 ft/sec² and the car at B is traveling at 30 ft/sec with acceleration of 5 ft/sec². Determine the relative velocity and relative acceleration of A and their directions with respect to B at this moment.

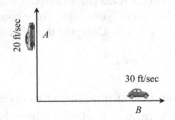

FIGURE 2.48   Positions of Two Cars for Problem 2.11

Problem 2.22
At the instant shown in Figure 2.49, the car at *A* is traveling at 10 m/sec with acceleration of 3 m/sec². The car at *B* is traveling at 14 m/sec with acceleration of 2 m/sec². Determine the relative velocity and relative acceleration of *B* and their directions with respect to *A* at this moment.

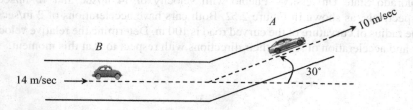

FIGURE 2.49 Positions of Two Cars for Problem 2.22

Problem 2.23
At the instant shown in Figure 2.50, the car at *A* is traveling at 10 m/sec with acceleration of 4 m/sec². The car at *B* is traveling at 15 m/sec with acceleration of 5 m/sec². Determine the relative velocity and relative acceleration of *A* and their directions with respect to *B* at this moment.

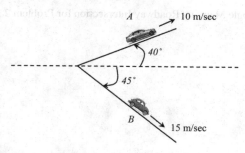

FIGURE 2.50 Positions of Two Cars for Problem 2.23

Problem 2.24
Two cars, *A* and *B* are running in two different roadways with velocity of 14 m/sec and 10 m/sec, respectively, as shown in Figure 2.51. Both cars have accelerations of 2 m/sec². The radius of curvature of the curved road is 100 m. Determine

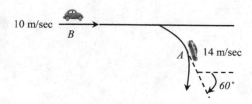

FIGURE 2.51 Positions of Two Cars for Problem 2.24

the relative velocity and acceleration of A and their directions with respect to B at this moment.

Problem 2.25

Two cars, A and B, are running in two different roadways at US-50 close to Colorado State University—Pueblo with velocity of 14 m/sec and 10 m/sec, respectively, as shown in Figure 2.52. Both cars have accelerations of 2 m/sec². The radius of curvature of the curved road is 100 m. Determine the relative velocity and acceleration of A and their directions with respect to B at this moment.

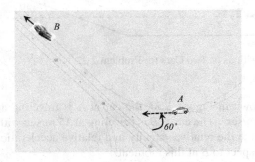

FIGURE 2.52 Google Map of a Roadway Intersection for Problem 2.25

3 Kinetics of Particles

3.1 GENERAL

Three laws discovered by scientist Sir Isaac Newton are the milestone of mechanics. Many problems in engineering have been successfully solved by applying these laws. Explaining linear and rotational motion, momentum, conservation principle, etc., are some of the successes of Newtonian mechanics. Particle kinetics is the study of the movement of particles and the forces that cause this movement.

3.2 CONCEPT OF FORCE

If you kick a soccer ball, it usually moves ahead. By hitting a steel ball of the same size, it is impossible to make a similar motion. A jet plane cannot be moved by pushing alone; however, a bike moving at a particular speed can be stopped by pulling it from behind. From this explanation, it is understood that the body having a large mass cannot change its inertia of rest or inertia of motion easily. The larger the mass, the higher the inertia. It can be seen from the above examples that in order to change the state of a body at rest or a body in motion, some external force must be applied. In our daily lives, we often push bodies to change their states. Sometimes, we move items from one place to another by pulling or lifting. In all cases, these need to have physical contact between the object and the person who applies the force. This type of force is called contact force. Examples of contact or tangential force are frictional force, a force created due to collision, stretched force, etc. It can be said that whatever is used to change the state of rest or state of motion of a body is called force.

Many events are happening in nature where two bodies are attracting or repelling each other. They may be attracted to each other without being in close proximity or even touching. For example, it is not easy to know whether a mango on a tree is attracting another mango on the same tree or if the mango is being attracted by the Earth. When the mango falls to the ground, it lands due to either the Earth's attraction or because of its weight. This type of force of attraction is called gravitational force. When a box is pulled along the floor, there is a force acting between the floor and the box, which resists the box's motion. This retarding force is called frictional force. Nucleons exist closely or side by side inside the nucleus of the atom. In this case, there is a force of attraction for which the nucleons do not get separated. This type of attraction in the nucleus is called the nuclear force.

In practice, there is not a single body on which force from outside does not act. If the resultant of two or more forces acting on a body from the outside is zero, then no effect of forces is observed. For example, if you stand on a hard concrete floor, the force applied by your body mass on the concrete floor and the reaction of

DOI: 10.1201/9781003283959-3

the concrete floor on you are equal; thus, no sinking of the floor occurs. However, if you stand on soft mud near a river, the mud cannot apply an equal reactive force to your weight. Therefore, you keep sinking until your feet touch the hard soil underneath. These action and reaction forces are called contact forces.

From Newton's First Law of Motion, we can get the idea of force. If no force is applied to a body, the body would stay in its state of rest indefinitely; likewise, if there is no intervention by a force, a moving body would likely move indefinitely in a straight line. This property of the body is called inertia.

3.3 TYPES OF FORCES

Although we are familiar with different kinds of forces in nature and these forces have different names, all forces are not fundamental. The fundamental or natural forces are those that are not derived from other forces; rather, these other forces are derived from the fundamental forces. In nature there are four fundamental forces. Other forces can be explained by one or more of the fundamental forces. These fundamental forces are:

(a) Gravitational Force – the attraction force between any two particles or objects, including the Earth.
(b) Electromagnetic Force – the force between two charged particles and between two magnetic materials.
(c) Strong Nuclear Force – the strong attraction force that keeps the nucleus stable in an atom.
(d) Weak Nuclear Force – the weak force that emits neutrino and β-particles from radioactive nuclei such as uranium, thorium, etc.

3.4 NEWTON'S SECOND LAW OF MOTION

We get the qualitative idea about Newton's First Law of Motion from the definition of force. Newton's Second Law enables us to measure that force. Before studying the second law, we should know about momentum. Suppose there is a collision between two cars. After the collision, how will you ascertain the direction of the moving bodies? By their velocities? By their masses? Or imagine you need to stop a bike that is rolling toward you at 5 mph versus your need to stop a car that is moving toward you at 5 mph. Obliviously, you will need to apply more force for stopping the car, although the speeds are equal in both cases. The reason for these events is momentum. It can be said that the motion produced in a body due to the combined effect of mass and velocity is momentum. While keeping the mass constant, if the velocity increases, the momentum of the body also increases. That means if the same body moves at a higher speed, then the momentum also becomes higher. Thus, the momentum (P), the product of mass, and its velocity can be expressed as:

$$P = mv$$

where m is the mass of the body and v is the speed. Its unit is kg.m/sec or lb.ft/sec.

Newton's Second Law states that the rate of change of momentum of a body is proportional to the applied force on the body and the change of momentum occurs along the direction of the force. Mathematically, it can be expressed as:

$$F \propto \frac{dP}{dt}$$

$$F = k\frac{dP}{dt} \qquad \left[\text{where } k \text{ is a constant of proportionality}\right]$$

$$F = k\frac{d}{dt}(mv)$$

$$F = km\frac{d}{dt}(v)$$

$$F = kma$$

From the definition of force, when $m = 1$ kg, $a = 1$ m/sec^2, then $F = 1$ N. From this definition, $k = 1$. So, the above equation can be written as:

$$F = m\frac{d}{dt}(v)$$

$$F = ma$$

Therefore, it can be said that the influence from outside, for which the state of a body, either at rest or in motion, is changed, is called force. It is measured by the product of mass (m) and acceleration (a), as shown below.

$$F = ma$$

The SI unit of force is N or kg.m/sec^2. The USCS unit is lbf or lb. It is noted that the definition of force conveys the meaning of Newton's First Law of Motion. The concept of force (F), mass (m), acceleration (a), and velocity (v) can be described as shown in Figure 3.1.

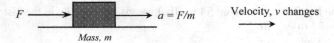

FIGURE 3.1   Relationship of Force and Acceleration

The vector form of Newton's Second Law of motion can be expressed as follows:

$$\vec{F} = m\vec{a}$$

Note that force, acceleration, velocity, etc., are vector quantities. If any force (F) acts on a body of mass (m), it attains some acceleration (a), and its velocity (v) changes with time (t). If there is no force (F), there is no acceleration (a), and the velocity (v) remains the same. If there is no force (F) at the beginning, there is no acceleration (a), and the velocity (v) is zero, meaning the body is at rest. If force (F) acts for some time, acceleration (a) is attained during this time, and the velocity (v) changes. If the force stops, the body keeps moving at a velocity attained at the last moment of the force application. Now practice the following worked-out examples for better understanding.

Example 3.1

A body was at rest. A force of 15 N acted on it for 4 sec, and then that force stopped. The body then moved 54 m in the next 9 sec. Calculate the mass of the body.

SOLUTION

Before starting the problem, recall that $F = ma$. This means if $F = 0$, $a = 0$. Similarly, if there is some force (F), there is some acceleration (a).

The question is asking for the mass (m) of the body. The force (F) at the first 4 sec is known. That means if we can find out the acceleration (a) in the first 4 sec, then the mass (m) can be determined.

First, let us understand the scenario of the example in Figure 3.2. A body was initially at rest, meaning the initial velocity $v_o = 0$. Then, a force of 15 N acted on it for 4 sec, and then the force stopped. That means the body attained some acceleration (a) during the first 4 sec and also attained some velocity during that time. Then, the force stopped. That means that after 4 sec, it had no acceleration as $F = 0$, and it moved at a uniform velocity (that it attained after the first 4 sec) in the last 9 sec.

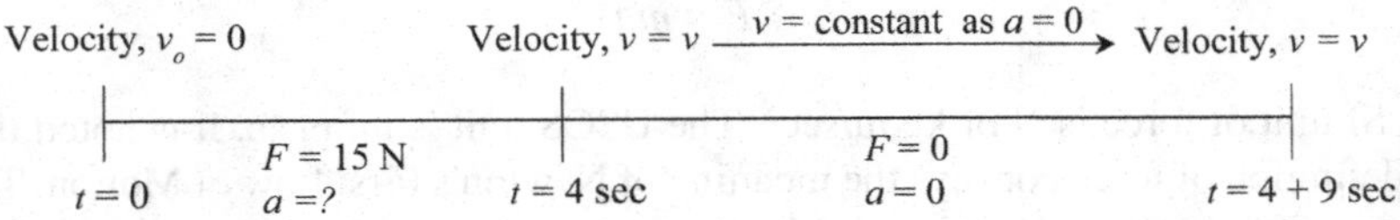

Velocity, $v_o = 0$ Velocity, $v = v$ $\underrightarrow{v = \text{constant as } a = 0}$ Velocity, $v = v$

$F = 15\,N$		$F = 0$
$t = 0$ $a = ?$	$t = 4\,sec$	$a = 0$ $t = 4 + 9\,sec$

FIGURE 3.2 Schematics of Example 3.1

In the last 9 sec, it traveled 54 m. That means the velocity in the last 9 sec,

$$v = \frac{s}{t_2} = \frac{54\,m}{9\,sec} = 6\,\frac{m}{sec}$$

Velocity after the first 4 sec = Velocity in the next 9 sec = 6 m/sec.
Since, at the beginning, the body was at rest, $v_o = 0$
Now, $v = v_o + a_o t_1$

$$6\,\frac{m}{sec} = 0 + a_o\left(4\,sec\right)$$

$$a_o = 1.5 \frac{m}{sec^2}$$

Therefore, $m = \dfrac{F}{a_o} = \dfrac{15 \text{ N}}{1.5 \dfrac{m}{sec^2}} = 10 \text{ kg}$

Answer: 10 kg

Example 3.2

A body acquires an acceleration of 3 m/sec² when a force of 7 N is applied to it. Determine the mass of the body. If the body is acted on by a force of 5 N along with 7 N force at an angle of 60°, determine the new acceleration of the body.

SOLUTION

Given:

  Force, $F = 7$ N
  Acceleration, $a_0 = 3$ m/sec²
  Mass, $m = ?$

Known: $F = ma_o$

  Therefore, $m = \dfrac{F}{a_o} = \dfrac{7 \text{ N}}{3 \dfrac{m}{sec^2}} = 2.33$ kg

If another force of 5 N along with the 7 N force acts on it, a new accelera-tion occurs due to the combined effect of these two forces. The resultant force (following the Parallelogram Method discussed in the previous chapter):

$$R = \sqrt{P^2 + Q^2 + 2PQ\cos\alpha} = \sqrt{(7N)^2 + (5N)^2 + 2(7N)(5N)\cos 60} = 10.44N$$

[Recall that if two forces P and Q act on a point with an angle of α, then the resultant force can be calculated using the Parallelogram Method as shown above.]

  Then, the new acceleration (a') can be calculated as:

$a' = \dfrac{R}{m} = \dfrac{10.44 \text{ N}}{2.33 \text{ kg}} = 4.48 \dfrac{m}{sec^2}$

Answers: 2.33 kg, 4.48 m/sec²

Example 3.3

A force of 10 N acts on a stationary body of mass of 2 kg. If the action of the force stops after 4 sec, then determine how far the body will travel in 8 sec from the start.

SOLUTION

Given,

Force, $F = 10$ N
Mass, $m = 2$ kg

Remember $F = ma$. If there is no force, then there is no acceleration. No acceleration means that the body is moving at a constant speed or at rest. A force of 10 N acts on the stationary body during the first 4 sec. That means the initial velocity $v_o = 0$, and there is some acceleration during the first 4 sec. Since the force stopped working on the body after 4 sec, it moved at a uniform velocity for the next 4 sec.

Acceleration in the first 4 sec: $a_o = \dfrac{F}{m} = \dfrac{10\,\text{N}}{2\,\text{kg}} = 5\,\dfrac{\text{m}}{\text{sec}^2}$

Travel in first 4 sec: $s_1 = v_o t + \dfrac{1}{2} a_o t^2 = 0(4\,\text{sec}) + \dfrac{1}{2}\left(5\,\dfrac{\text{m}}{\text{sec}^2}\right)(4\,\text{sec})^2 = 40\,\text{m}$

The velocity of the body after 4 sec: $v = v_o + a_o t = 0 + \left(5\,\dfrac{\text{m}}{\text{sec}^2}\right)(4\,\text{sec}) = 20\,\dfrac{\text{m}}{\text{sec}}$

Since there was no force after the first 4 sec, there is no acceleration, and the body moves at a constant speed, $v = 20$ m/sec.

$$s_2 = vt + \frac{1}{2} a_o t^2 = \left(20\,\frac{\text{m}}{\text{sec}}\right)(4\,\text{sec}) + \frac{1}{2}(0)t^2 = 80\,\text{m}$$

$$s = s_1 + s_2 = 40\,\text{m} + 80\,\text{m} = 120\text{ m}$$

Answer: 120 m

The above three examples show how the basic definitions of kinetics can be applied to solve such problems. All of the motions discussed so far are in a linear path on a plane. If motions are in three-dimensional (3D) space, vectors can be used to deal with such motions. Let us work on an example related to vectors. Knowledge of vectors (reviewed in the previous chapter) is required here as well. The following example will clarify the vector rule.

Example 3.4

Two forces, $\vec{F_1} = 5\vec{i} + 2\vec{j} - 2\vec{k}$ N and $\vec{F_2} = -3\vec{i} + \vec{j} - 7\vec{k}$ N, are applied on a 100-g particle. Determine the acceleration of the particle both in vector and scalar forms.

SOLUTION

Given,

Forces: $\vec{F_1} = 5\vec{i} + 2\vec{j} - 2\vec{k}$ N and $\vec{F_2} = -3\vec{i} + \vec{j} - 7\vec{k}$ N
Mass, $m = 100$ g $= 0.1$ kg

There are two forces acting on the particle. Acceleration will be generated due to the combined effect of these two forces. Remember that force, acceleration, velocity, etc., are vector quantities. Therefore, let the acceleration be $a_x\vec{i} + a_y\vec{j} + a_z\vec{k}$ where a_x, a_y, and a_z are the components of acceleration along the x, y, and z axes, respectively. $\vec{i}, \vec{j}$ and $\vec{k}$ are the unit vectors along the x, y, and z axes, respectively.

$$\sum \vec{F} = m\vec{a}$$

$$\vec{F_1} + \vec{F_2} = m\vec{a}$$

$$\left(5\vec{i} + 2\vec{j} - 2\vec{k}\right) + \left(-3\vec{i} + \vec{j} - 7\vec{k}\right) = (0.1\text{kg})\left(a_x\vec{i} + a_y\vec{j} + a_z\vec{k}\right)$$

$$2\vec{i} + 3\vec{j} - 9\vec{k} = 0.1a_x\vec{i} + 0.1a_y\vec{j} + 0.1a_z\vec{k}$$

Equating the corresponding unit vectors components

For $\vec{i}$: $0.1a_x = 2$ $\therefore a_x = 20 \dfrac{m}{\sec^2}$

For $\vec{j}$: $0.1a_y = 3$ $\therefore a_y = 30 \dfrac{m}{\sec^2}$

For $\vec{k}$: $0.1a_z = -9$ $\therefore a_z = -90 \dfrac{m}{\sec^2}$

Therefore, the acceleration in vector form, $\vec{a} = 20\vec{i} + 30\vec{j} - 90\vec{k}$
The   acceleration   in   scalar,

$$|a| = \sqrt{a_x{}^2 + a_y{}^2 + a_z{}^2} = \sqrt{\left(20\dfrac{m}{\sec^2}\right)^2 + \left(30\dfrac{m}{\sec^2}\right)^2 + \left(-90\dfrac{m}{\sec^2}\right)^2} = 97\dfrac{m}{\sec^2}$$

Note: If vector form of the acceleration is not asked, then the solution can be simplified as follows:

Resultant Force, $\vec{F} = \vec{F_1} + \vec{F_2} = \left(5\vec{i} + 2\vec{j} - 2\vec{k}\right) + \left(-3\vec{i} + \vec{j} - 7\vec{k}\right) = 2\vec{i} + 3\vec{j} - 9\vec{k}$

$$|F| = \sqrt{F_x{}^2 + F_y{}^2 + F_z{}^2} = \sqrt{(2N)^2 + (3N)^2 + (-9N)^2} = 9.7N$$

$$a = \frac{F}{m} = \frac{9.7N}{0.1kg} = 97 \frac{m}{\sec^2}$$

Answer: $97 \dfrac{m}{\sec^2}$

The above example describes a motion in 3D space. The forces mentioned are not time-sensitive (i.e., do not change with time). That means the forces were constant. From the rule of $F = ma$, if F is constant, a is also constant. Now, we will work on an example that has time-dependent forces (i.e., time-dependent acceleration).

Example 3.5

Two forces, $\vec{F_1} = t\vec{i} + 2\vec{j} - 2\vec{k}$ and $\vec{F_2} = -3\vec{i} + \vec{j} - t\vec{k}$, are applied on a 6.44-lb particle at rest, where the forces are in lbf or lb and t is in sec. Determine the following:

(a) The acceleration of the particle after 2 sec
(b) The velocity of the particle after 2 sec
(c) The displacement in 2 sec

SOLUTION

Given,

Forces: $\vec{F_1} = t\vec{i} + 2\vec{j} - 2\vec{k}$ and $\vec{F_2} = -3\vec{i} + \vec{j} - t\vec{k}$

Weight = 6.44 lb

Mass, $m = \dfrac{6.44 \text{ lb}}{32.2 \dfrac{\text{ft}}{\text{sec}^2}} = 0.2 \text{ slug}$

(a) Let the acceleration be $a_x\vec{i} + a_y\vec{j} + a_z\vec{k}$.

$$\sum \vec{F} = m\vec{a}$$

$$\vec{F_1} + \vec{F_2} = m\vec{a}$$

$$\left(t\vec{i} + 2\vec{j} - 2\vec{k}\right) + \left(-3\vec{i} + \vec{j} - t\vec{k}\right) = \left(0.2\,\text{slug}\right)\left(a_x\vec{i} + a_y\vec{j} + a_z\vec{k}\right)$$

$$(t-3)\vec{i} + 3\vec{j} - (t+2)\vec{k} = 0.2a_x\vec{i} + 0.2a_y\vec{j} + 0.2a_z\vec{k}$$

Equating the corresponding components:
    For $\vec{i}$: $0.2a_x = t - 3$

$$a_x = 5t - 15 \ \frac{\text{ft}}{\text{sec}^2}$$

$$a_x\big|_{t=2\sec} = 5(2\sec) - 15 = -5 \ \frac{\text{ft}}{\text{sec}^2}$$

For $\vec{j}$: $0.2a_y = 3$

$$a_y = 15 \frac{ft}{sec^2}$$

$$a_y\big|_{t=2sec} = 15 \frac{ft}{sec^2}$$

For $\vec{k}: 0.2a_z = -(t+2)$

$$a_z = -5t - 10 \frac{ft}{sec^2}$$

$$a_z\big|_{t=2sec} = -5(2sec) - 10 = -20 \frac{ft}{sec^2}$$

The acceleration in vector form, $\vec{a} = -5\vec{i} + 15\vec{j} - 20\vec{k}$

The acceleration in scalar, $|a| = \sqrt{a_x^2 + a_y^2 + a_z^2}$

$$= \sqrt{\left(-5\frac{ft}{sec^2}\right)^2 + \left(15\frac{ft}{sec^2}\right)^2 + \left(-20\frac{ft}{sec^2}\right)^2} = 25.5 \frac{ft}{sec^2}$$

(b) Velocity after 2 sec (Recall $a = \dfrac{dv}{dt}$ from the previous chapter)

$$a_x = 5t - 15$$

$$a_x = \frac{dv_x}{dt}$$

$$v_x = \int dv_x = \int_0^{2sec} a_x dt = \int_0^{2sec} (5t-15)dt = \left[\frac{5t^2}{2} - 15t + C\right]_0^2 = -20 \frac{ft}{sec}$$

$$v_y = \int dv_y = \int_0^{2sec} a_y dt = \int_0^{2sec} 15 dt = [15t]_0^2 = 30 \frac{ft}{sec}$$

$$v_z = \int dv_z = \int_0^{2sec} a_z dt = \int_0^{2sec} (-5t-10)dt = \left[-\frac{5t^2}{2} - 10t + C\right]_0^2 = -30 \frac{ft}{sec}$$

$$|v| = \sqrt{v_x^2 + v_y^2 + v_z^2} = \sqrt{\left(-20\frac{ft}{sec}\right)^2 + \left(30\frac{ft}{sec}\right)^2 + \left(-30\frac{ft}{sec}\right)^2} = 46.9 \frac{ft}{sec}$$

(c) Displacement in 2 sec (Recall $v = \dfrac{ds}{dt}$ from the previous chapter)

$$v_x = \frac{5t^2}{2} - 15t$$

$$s_x = \int_0^{2\sec} v_x dt = \int_0^{2\sec} \left(\frac{5t^2}{2} - 15t \right) dt = \left[\frac{5t^3}{6} - \frac{15t^2}{2} + C \right]_0^2 = -23.3 \text{ ft}$$

$$s_y = \int_0^{2\sec} v_y dt = \int_0^{2\sec} (15t) dt = \left[\frac{15}{2} t^2 + C \right]_0^2 = 30 \text{ ft}$$

$$s_z = \int_0^{2\sec} v_z dt = \int_0^{2\sec} \left(-\frac{5t^2}{2} - 10t \right) dt = \left[-\frac{5t^3}{6} - \frac{10t^2}{2} + C \right]_0^2 = -26.7 \text{ ft}$$

$$|s| = \sqrt{s_x^2 + s_y^2 + s_z^2} = \sqrt{(-23.3 \text{ ft})^2 + (30 \text{ ft})^2 + (-26.7 \text{ ft})^2} = 46.4 \text{ ft}$$

Answers: 25.5 ft/sec², 46.9 ft/sec, 46.4 ft

3.5   APPLICATIONS OF NEWTON'S LAW OF MOTION

When a body applies force to another body, then the second body also applies an equal and opposite force to the first body. We learned about these action and reaction forces from Newton's Third Law of Motion. In nature, forces act in pairs, not as separate individual forces. That means two forces are complementary to each other. One of these forces is called the action force, and the other is called the reaction force. As long as the action force is there, the reaction force also exists. Some practical applications of Newton's laws of motion are given below:

(a) *Driving a Car.* While driving a car, a mechanical forward force is applied first, which can be viewed as a pulling force. The friction between the wheels and the road's surface applies a force opposite to the impending motion. When the applied pulling force (F_P) is greater than the friction force (F_f), the car starts running and attains some acceleration (a). This procedure can be schematically shown as in Figure 3.3:

   Mathematically, it can be expressed as:

$$\sum F = F_P - F_f = ma$$

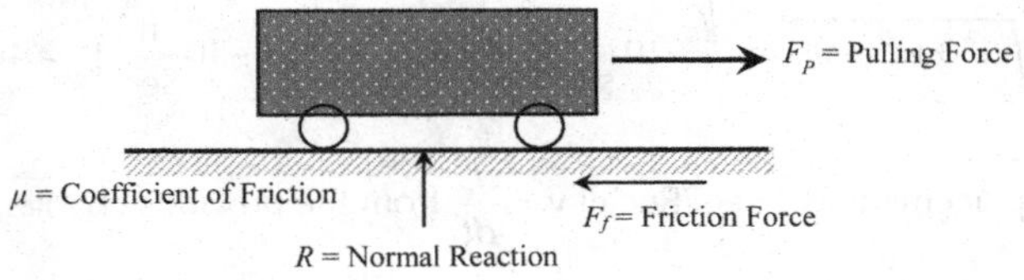

F_P = Pulling Force

μ = Coefficient of Friction

F_f = Friction Force

R = Normal Reaction

FIGURE 3.3   Driving a Car on a Roadway

where:
m = The mass of the car
a = The resulting acceleration

From statics, we learned that $F_f = \mu R$. This means that the frictional force is the product of the coefficient of friction and the normal reaction of the body. The frictional force always works opposite to the motion of the body.

(b) *Firing of a Bullet from a Gun.* When a bullet is fired from a gun, it moves ahead with tremendous speed. If the gun applies a force, F, on the bullet, the bullet also applies an equal and opposite force on the gun. Due to this reaction, the gun recoils backward, which can be explained by momentum. Before firing, both the gun and the bullet remain at rest. So, the momentum of both the gun and the bullet is zero. Hence their total initial momentum is zero. After firing, the bullet moves ahead with a velocity due to the explosion and thus attains forward momentum. Now according to the conservation of momentum principle, the total momentum after firing will be zero. Therefore, the gun will acquire an equal and opposite momentum, thus, having a backward motion.

Suppose a bullet of mass, m, is released with a velocity, v, from a gun with a mass, M. Again, suppose the velocity of the gun after firing = V.

Before firing the total momentum = momentum of the gun

+momentum of the bullet

$$= M(0) + m(0)$$

$$= 0$$

After firing the momentum = momentum of the gun + momentum of the bullet

$$= MV + mv$$

However, according to the conservation principle, momentum before and after must be equal. Therefore:

The total momentum before firing = The total momentum after firing

$$0 = MV + mv$$

$$MV = -mv$$

In other words, the momentum of the gun = the momentum of the bullet (opposite to each other).

Example 3.6

A 100-kg block resting on a horizontal surface is being pulled with a force of 2,000 N. The coefficient of kinetic friction between the block and the plane is 0.2. Determine the velocity of the block after traveling 10 m.

SOLUTION

Given,

Mass, m = 100 kg
Pulled force, F_{pulled} = 2,000 N
Coefficient of kinetic friction = 0.2
Distance traveled, s = 10 m
Velocity after 10 m, v = ?

We know that there are two equations to solve for velocity: $v^2 = v_o^2 + 2a_o s$ and $v = v_o + a_o t$. This problem does not say anything about time (t). Thus, the first equation ($v^2 = v_o^2 + 2a_o s$) may be appropriate. However, the acceleration (a_o) is still unknown. As the mass (m) and information about force (F) are given, acceleration (a_o) can be worked out.

Let us work to find the acceleration (a_o) first. We know, $\sum F = ma_o$

Therefore, $\sum (F_{applied} - F_{friction}) = ma_o$

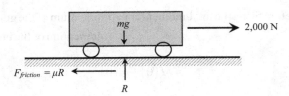

FIGURE 3.4 The Forces on the Body

From Figure 3.4
$\sum F$ along the vertical direction = 0
$R - mg = 0$
$R = mg$
$F_{friction} = \mu R$
$= \mu (mg)$
$= 0.2(100 \text{ kg})(9.81 \text{ m/sec}^2)$
$= 196.2 \text{ N}$

$$\sum (2,000 \text{N} - 196.2 \text{N}) = 100 \text{kg} (a)$$

Therefore, a_o = 18 m/sec²
Velocity after traveling 10 m:

$$v^2 = v_o^2 + 2a_o s$$

$v^2 = 0 + 2$ (18 m/sec²) (10 m)
$v^2 = 360$ m²/sec²
$v = 18.98$ m/sec
$v \approx 19$ m/sec

Answer: 19 m/sec

Example 3.7

A bullet is fired at a speed of 180 cm/sec from a gun with a mass of 20 kg. If the mass of the bullet is 30 g, find out the velocity of the gun.

SOLUTION

Given:

Velocity of the bullet, $v = 180$ cm/sec $= 1.8$ m/sec
Mass of the gun, $M = 20$ kg
Mass of the bullet, $m = 30$ g $= 0.03$ kg
Velocity of the gun, $V = ?$

We know: Momentum of the gun $= -$ Momentum of the bullet

$$MV = -mv$$

$$V = -\frac{mv}{M} = -\frac{(0.03 \text{ kg})\left(1.8 \dfrac{m}{sec}\right)}{20 \text{ kg}} = -0.0027 \frac{m}{sec}$$

Answer: 0.0027 m/sec (opposite to the bullet)

3.6   WORK AND ENERGY OF PARTICLES

3.6.1   WORK

In general, doing something is called work; for example, to study, work in factories, ride a bike, etc. However, in the language of science, work has a different meaning. If there is a displacement of a body due to the application of force, only then is work done. For example, a book is allowed to fall from a table to the ground. A person reached the top of a hill step by step carrying his backpack. These are two examples of work. In the first case, the displacement has been along the direction of the gravitational force, and in the second case, the displacement has been against the gravitational force. So, work has been done in both cases. A person standing at a fixed place carrying a load on his shoulder has become very tired, but no work is done in this case as there is no displacement. From this discussion, it is understood that if there is displacement due to the application of force, only then is work done. If there is no displacement, even if force is applied, no work will be done. The unit of work is N.m or J. This means, 1 J = 1 N.m. In the US customary unit, the unit of work is lb.ft.

In our daily lives, we see many examples of work around us. For example, kicking a soccer ball, pushing a shopping cart, driving cars, climbing up a mountain, etc., are all examples of work. In summary, if a force is applied to a body, and displacement occurs, then the product of the force and the component of displacement along the direction of the force is called work.

Figure 3.5 shows a force along the vertical axis and the displacement along the horizontal axis. The applied force here is constant and acting in the horizontal direction to the right. Due to the application of this force, the body moves from the position, x_1 to the position, x_2. Therefore, the displacement is $(x_2 - x_1)$. The area of the rectangle represents the total work done.

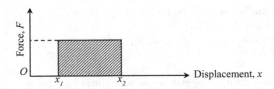

FIGURE 3.5   Concept of Work for Constant Force

If the applied force is variable, then the graph can be of any shape, as shown in Figure 3.6. In a small strip, dx, the force, $F(x)$, can be assumed constant. For this small strip, the work done is $F(x)dx$. The total work can be determined as:

$$W = U_{1\to2} = F_1 dx_1 + F_2 dx_2 + \dots$$

$$W = U_{1\to2} = \int_{x_1}^{x_2} F(x)\, dx$$

Where F_1, F_2, etc., are forces in segments, Δx_1, Δx_2, etc., respectively. $U_{1\to2}$ is often used to denote the net work done (W) on a particle when moving from position 1 to position 2.

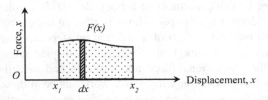

FIGURE 3.6   Concept of Work for Variable Force

If the displacement occurs along the direction making an angle θ, as shown in Figure 3.7, with the applied force, then work is equal to the product of the force and the component of displacement along the direction of the force. Mathematically, it can be shown as follows:

$$W = Fs \cos\theta$$

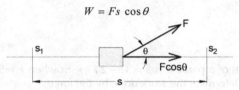

FIGURE 3.7   Work along a Straight Line for a Constant Force

In vector form, the work can be presented as the dot product of $\vec{F}$ and $\vec{s}$ as follows:

$$W = \vec{F} \cdot \vec{s}$$

Knowledge of dot product of vectors is required to apply the above equation. Consider two vectors: $\vec{A} = a_x\vec{i} + a_y\vec{j} + a_z\vec{k}$ and $\vec{B} = b_x\vec{i} + b_y\vec{j} + b_z\vec{k}$. The dot product of two vectors is a scalar quantity and can be determined as:

$$\vec{A}.\vec{B} = (a_xb_x) + (a_yb_y) + (a_zb_z)$$

The following example will clarify this vector rule.

Example 3.8

A constant force, $\vec{F} = \left(-\vec{i} + 2\vec{j} + 3\vec{k}\right)$ N, acts on a body. If the displacement is 3 m along the z-axis, then determine the work done on the body.

SOLUTION

Given:

Force, $\vec{F} = \left(-\vec{i} + 2\vec{j} + 3\vec{k}\right)$

Displacement, $\vec{s} = 3\vec{k} = 0\vec{i} + 0\vec{j} + 3\vec{k}$

Known: $W = \vec{F} \cdot \vec{s}$

Work: $W = \vec{F} \cdot \vec{s} = \left(-\vec{i} + 2\vec{j} + 3\vec{k}\right) \cdot \left(0\vec{i} + 0\vec{j} + 3\vec{k}\right)$

$$= (-1)(0) + (2)(0) + (3)(3)$$
$$= 0 + 0 + 9$$
$$= 9\,J$$

Answer: 9 J

Example 3.9

The displacement of a particle is $\vec{s} = 3\vec{i} - 2\vec{j} + \vec{k}$ m when $\vec{F} = 5\vec{i} + 3\vec{j} - 2\vec{k}$ N is applied to it. Determine the work done by the force.

SOLUTION

Given:

Force, $\vec{F} = 5\vec{i} + 3\vec{j} - 2\vec{k}$

Displacement, $\vec{s} = 3\vec{i} - 2\vec{j} + \vec{k}$

Known: $W = \vec{F} \cdot \vec{s} = (5\vec{i} + 3\vec{j} - 2\vec{k}) \cdot (3\vec{i} - 2\vec{j} + \vec{k})$

$$= (5)(3) + (3)(-2) + (-2)(1)$$

$$= 15 - 6 - 2$$
$$= 7\,J$$

Answer: 7 J

3.6.2 ENERGY

3.6.2.1 Definitions

When a body works, it is considered to have energy. Energy is measured by the total amount of work that a body can do. That means, when work is done, we can measure energy. The SI unit of energy is J (N.m), and the US customary unit is lb.ft. The units for energy and work are the same. When a body does work, its energy decreases, and the body on which work is done has its energy increased. The body which has less capacity to do work has less energy. It can be said that work is the measure of energy. Energy has different forms:

- Mechanical energy
- Heat energy
- Light energy
- Sound energy

- Magnetic energy
- Electrical energy
- Chemical energy
- Nuclear energy
- Solar energy, etc.

Energy is present in different forms. Different forms of energy are related to one another. It is possible to transform from one form of energy to another form, and this is called the transformation of energy. A few examples of the transformation of energy are given below:

(a) Water flows from a higher to a lower elevation. Its energy at the higher places is potential energy. While flowing downward, its potential energy changes to kinetic energy. This kinetic energy can be used to generate electricity by using rotating turbine shafts. That means the mechanical energy is being transformed into electrical energy.
(b) When electric energy passes through an electric bulb, we get light. Hence, electric energy is being transformed into light energy.
(c) When current passes through an electric iron, heat is produced. We iron clothes by using this heat. Here electric energy is being converted into thermal energy.

When energy is transferred from one form to another, no loss or gain of energy occurs. That means it is impossible to create or destroy energy. When one form of energy disappears, it reappears in another form. This is called the conservation of energy.

3.6.2.2 Kinetic Energy (T)

When a nail is struck by a hammer on a wall, the nail enters overcoming the resistance of the wall. The hammer can do this work because of its motion and the resulting kinetic energy, which allows the nail to overcome the resistance of the wall. You might have seen sailboats on the river. The kinetic energy of the current in the river drives the boat. When a strong wind blows, the boat moves faster if the sail is hoisted. As another example, an athlete would not jump from rest when competing in the high jump or long jump but would instead jump after running a certain distance. As a result, they can go a longer distance by using this kinetic energy. From this discussion, we observe that to stop a moving body by applying an external force, the total work done by the body before stopping is the measure of the kinetic energy of the body. In short, the energy possessed by a body by virtue of its motion is called kinetic energy. In other words, if a particle is at rest, a certain amount of work is required to be applied to it in order to produce a certain velocity. This work is also known as kinetic energy. The SI unit of kinetic energy is also J (N.m), and the US customary unit is lb.ft.

Suppose a body of mass, m, is moving along AB with a velocity of v, as shown in Figure 3.8. A constant force F is applied opposite to the motion along with BA.

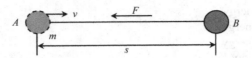

FIGURE 3.8 Concept of Kinetic Energy

Due to this, uniform deceleration will be produced. Let the uniform deceleration $= a$, and the body came to rest at point B after covering a distance s from the point, A. Hence the final velocity is $v = 0$.

> Kinetic energy = work done before coming to rest
> = Force × Distance traveled before coming to rest
> = $F(s)$

From Newton's Second Law of Motion, $F = ma_o$

Again, $v^2 = v_o^2 + 2a_o s$

$$0 = v^2 - 2a_o s$$

$$s = \frac{v^2}{2a_o}$$

$$\text{Kinetic energy} = T = Fs = \left(ma_o\right)\left(\frac{v^2}{2a_o}\right) = \frac{1}{2}mv^2$$

This means the particle's initial kinetic energy plus the work done by all the forces acting on the particle as it moves from its initial to its final position equals the final kinetic energy of the particle.

3.6.2.3 Work-Energy Theorem

Let a force F be applied on an object of mass, m, and the velocity of the object changes from v_1 to v_2 when the object travels a distance, s.

$$v_2^2 = v_1^2 + 2a_o s$$

$$s = \frac{v_2^2 - v_1^2}{2a}$$

Therefore, the net work done by the force,

$$W = Fs = \left(ma\right)\left(\frac{v_2^2 - v_1^2}{2a}\right) = \frac{1}{2}m\left(v_2^2 - v_1^2\right) = \frac{1}{2}mv_2^2 - \frac{1}{2}mv_1^2$$

$$W = \frac{1}{2}mv_2^2 - \frac{1}{2}mv_1^2$$

$$U_{1\rightarrow 2} = \frac{1}{2}mv_2^2 - \frac{1}{2}mv_1^2$$

$$U_{1\rightarrow 2} = T_2 - T_1$$

$$T_1 + U_{1\rightarrow 2} = T_2$$

i.e., Initial Kinetic Energy + Net Work Done = Final Kinetic Energy

Hence, the increase in kinetic energy of an object is equal to the work done by the applied force. This is called the work-energy theorem. Sometimes people forget that the work-energy theorem only applies to the net work, not the work done by a single force. The work-energy theorem states that the net work done by the forces on an object equals the change in its kinetic energy.

Example 3.10

A 100-kg block resting on a horizontal surface is being pulled at a force of 2,000 N. The coefficient of kinetic friction between the block and the surface is 0.2. Determine the velocity of the block after traveling 10 m.

SOLUTION

This is the same example as Example 3.6. However, this time it will be solved using the work-energy theorem.

Given,

Initial velocity, $v_o = 0$
Mass, $m = 100$ kg
Pulled force, $F_{pulled} = 2,000$ N
Coefficient of kinetic friction = 0.2
Distance traveled, $s = 10$ m
Velocity after 10 m, $v = ?$
We know: $T_1 + U_{1\rightarrow 2} = T_2$

$$T_2 = T_1 + U_{1\rightarrow 2}$$

$$\frac{1}{2}mv_2^2 = \frac{1}{2}mv_1^2 + U_{1\rightarrow 2}$$

Initial velocity, $v_o = 0$, means the initial kinetic energy $\frac{1}{2}mv_1^2 = \frac{1}{2}m(0)^2 = 0$.

Net work done on the body, $U_{1\rightarrow 2} = \sum F.s = (F_{applied} - F_{friction})s$

$F_{friction} = \mu R = \mu(mg) = 0.2(100 \text{ kg})(9.81 \text{ m/sec}^2) = 196.2$ N
Net work done on the block:

$$U_{1\rightarrow 2} = \sum F.s = (2,000\text{N} - 196.2\text{N})10\text{m} = 18,038 \text{ J}$$

Now, $\frac{1}{2}mv_2^2 = \frac{1}{2}m(0)^2 + 18,038\text{J}$

$$v_2 = 18.99 \frac{m}{sec}$$

Therefore, $v_2 \approx 19 \frac{m}{sec}$

Answer: 19 m/sec

Example 3.11

A bullet can penetrate a piece of wood. If the velocity of the bullet increases to three times, determine how many pieces of wood of the same thickness can be penetrated.

SOLUTION

Let,

The mass of the bullet = m
The thickness of 1 piece of wood = x
1st-time initial velocity = v
2nd-time initial velocity = $3v$

1st case:

$$T_1 + U_{1\rightarrow2} = T_2$$

$$\frac{1}{2}mv_1^2 + U_{1\rightarrow2} = \frac{1}{2}mv_2^2$$

$$\frac{1}{2}mv^2 + (-F_f)(x) = \frac{1}{2}m(0)^2$$

$$F_f x = \frac{1}{2}mv^2$$

2nd case:

$$T_1 + U_{1\rightarrow2} = T_2$$

$$\frac{1}{2}mv_1^2 + U_{1\rightarrow2} = \frac{1}{2}mv_2^2$$

$$\frac{1}{2}m(3v)^2 + (-F_f)(nx) = \frac{1}{2}m(0)^2$$

$$n = \frac{\frac{1}{2}m(3v)^2}{F_f x} = \frac{\frac{1}{2}m(3v)^2}{\frac{1}{2}m(v)^2} = 9$$

Answer: 9

Example 3.12

A body of 10 kg is dropped from a height of 100 m. Determine its kinetic energy when it is 60 m above the ground.

SOLUTION

Given,

> Mass, $m = 10$ kg
> Travel distance, $y = 100$ m $- 60$ m $= 40$ m [Figure 3.9]
> Initial velocity, $v_{oy} = 0$ m/sec

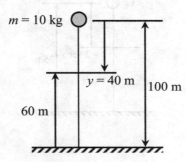

FIGURE 3.9

Velocity after falling 40 m (60 m above the ground)

$$v_y^2 = v_{0y}^2 + 2gy$$

$$v_y = \sqrt{0 + 2\left(9.81\frac{m}{sec^2}\right)(40\,m)}$$

$$v_y = 28.01\frac{m}{sec}$$

Therefore, the kinetic energy, $T = \frac{1}{2}mv_y^2$

$$= \frac{1}{2}(10\,kg)\left(28.01\frac{m}{sec}\right)^2$$

$$= 3,923\,J$$

The kinetic energy when it is 60 m above the ground is 3,923 J.

Answer: 3,923 J

Example 3.13

A 4-kg mass is allowed to fall freely from a height of 25 m by gravitational attraction. After 2 sec, determine the kinetic energy of the mass.

SOLUTION

Given,

Mass, $m = 4$ kg
Travel time, $t = 2$ sec [Figure 3.10]

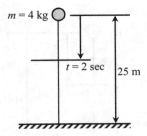

FIGURE 3.10

Kinetic energy after 2 sec, $T = \frac{1}{2} mv^2$. We know the mass, m, and we need to find out v.
Velocity after 2 sec, $v_y = v_{oy} + gt$
$v_y = 0 + 2g$
$v_y = 2g$
Kinetic energy after 2 sec, $T = \frac{1}{2} mv_y^2$
$= \frac{1}{2} m (2g)^2$
$= 2mg^2$
$= 2 (4 \text{ kg}) (9.81 \text{ m/sec}^2)^2$
$= 769.9$ J

Answer: 769.9 J

Example 3.14

A car with a mass of 2,000 kg, while descending at a velocity of 16 m/sec along a road inclined at 30° with the ground level, is brought to rest by the driver within 40 m by applying the brake. Determine how much opposing force was applied to the car.

SOLUTION

The car is running inclined downward. After applying the brake, the brake force (F_{brake}) and the friction force (F_f) help to stop the motion. Let the combined opposing force be $F_R = F_{brake} + F_f$ (upward). Here, the downward kinetic energy of the car is resisted by the upward opposing force. Consider the downward incline direction as positive and the upward incline direction as negative. The weight of the car, as shown in Figure 3.11, has been resolved into two components – one along the inclined surface downward ($W\sin\theta$), and the other one being normal to the inclined surface ($W\cos\theta$).

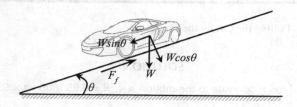

FIGURE 3.11   Resolution of the Weights of the Car in the Inclined Surface

Initial kinetic energy, $T_1 = \dfrac{1}{2}mv_1^2 = \frac{1}{2}(2{,}000 \text{ kg})(16 \text{ m/sec})^2 = 256{,}000 \text{ J}$

Net work done, $U_{1 \to 2} = \sum F.s = (-F_R + mg \sin 30°) (40 \text{ m})$

Final kinetic energy, $T_2 = \dfrac{1}{2}mv_2^2 = \frac{1}{2}(2{,}000 \text{ kg})(0)^2 = 0$

Now, $T_1 + U_{1 \to 2} = T_2$
$256{,}000 \text{ J} + (-F_R + mg \sin 30°) (40 \text{ m}) = 0$
$[F_R - (2{,}000 \text{ kg}) (9.81 \text{ m/sec}^2) \sin 30°] (40 \text{ m}) = 256{,}000 \text{ J}$
$F_R = 16{,}210 \text{ N}$

Answer: 16,210 N

Note: The resistive force here includes the friction force. As the friction force is not specified, the combined force, including the friction force and the brake force, has been calculated. If the question asks about the brake force only, then:

$F_{brake} = F_R - F_f$
$F_{brake} = 16{,}210 \text{ N} - \mu \, mg \cos\theta$
μ , friction coefficient, must be known to solve this.

Example 3.15

A 2,000-kg car is being pulled by a heavy truck, as shown in Figure 3.12. The tension developed in the towing boom is 15 kN. The net load on the rear axle of the car, considering the weight distribution and the pull, is 800 kg. When the travel distance is 150 m, the car moves ahead with a speed of 8 m/sec. Determine its speed when the travel distance is 250 m. The coefficient of kinetic friction between the tire and the ground is 0.25.

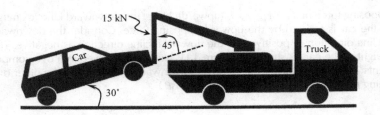

FIGURE 3.12   Pulling the Car by the Truck

SOLUTION

Frictional force opposite to the motion = $\mu_k R$ = 0.25 (800 kg × 9.81 m/sec²) = 1,962 N

Applied force toward the motion = 15,000 cos75° = 3,882 N

Initial kinetic energy, $T_1 = \frac{1}{2}mv_1^2 = \frac{1}{2}(800\,\text{kg})\left(8\,\frac{m}{\text{sec}}\right)^2 = 25,600\,\text{J}$

Final kinetic energy, $T_2 = \frac{1}{2}mv_2^2 = \frac{1}{2}(800\,\text{kg})v_2^2 = (400)v_2^2\,\text{J}$

Work done on the system, $U_{1\to2} = \sum Fs = (F - F_f)s$

$$= (3,882\,\text{N} - 1,962\,\text{N})(250\,\text{m} - 150\,\text{m})$$

$$= 192,000\,\text{J}$$

Now, $T_1 + U_{1\to2} = T_2$

$$T_2 = T_1 + U_{1\to2}$$

$$(400)v_2^2\,\text{J} = 25,600\,\text{J} + 192,000\,\text{J}$$

$$v_2 = 23.3\,\frac{m}{\text{sec}}$$

Answer: 23.3 m/sec

Example 3.16

When the driver applies the brakes of a car traveling at 45 mph, it skids 15 ft before stopping, as shown in Figure 3.13. Determine how far the car will skid if it is traveling at 80 mph when the brakes are applied.

FIGURE 3.13   Braking the Car for Example 3.16. Courtesy of Thompsons Solicitors

SOLUTION

The kinetic energy of the car will be dissipated by the resistive force. After applying the brake, the brake force (F_{brake}) and the friction force (F_f) help to stop the motion, as shown in Figure 3.14. Let the combined resistive force be F_R, and the mass of the car be m.

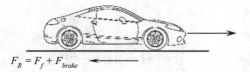

$$F_R = F_f + F_{brake}$$

FIGURE 3.14   Resistive Force in the Car for Example 3.16

First Case:

Initial velocity, $v_1 = 45 \dfrac{mile}{hour} = \dfrac{45\,mile\left(5,280\,\dfrac{ft}{mile}\right)}{(1h)\left(3,600\,\dfrac{sec}{h}\right)} = 66 \dfrac{ft}{sec}$

Final velocity, $v_2 = 0$
Travel distance, $s = 15$ ft
Resistive force, $F_R = ?$

$$T_1 + U_{1 \to 2} = T_2$$

$$\frac{1}{2}mv_1^2 + U_{1 \to 2} = \frac{1}{2}mv_2^2$$

$$\frac{1}{2}mv_1^2 + (-F_R)(15\,ft) = \frac{1}{2}m(0)^2$$

[F_R is negative since it is opposite to the direction of motion]

$$\frac{1}{2}m\left(66\frac{ft}{sec}\right)^2 - F_R(15ft) = \frac{1}{2}m(0)^2$$

$$F_R = 145.2m$$

Second Case:

Initial velocity, $v_1 = 80\,\dfrac{mile}{hour} = \dfrac{80\,mile\left(5,280\,\dfrac{ft}{mile}\right)}{(1h)\left(3,600\,\dfrac{sec}{h}\right)} = 117.3\,\dfrac{ft}{sec}$

Final velocity, $v_2 = 0$
Travel distance, $s = ?$

$$T_1 + U_{1\to2} = T_2$$

$$\frac{1}{2}mv_1^2 + U_{1\to2} = \frac{1}{2}mv_2^2$$

$$\frac{1}{2}mv_1^2 + (-F_R)(s) = \frac{1}{2}m(0)^2$$

$$\frac{1}{2}m\left(117.3\frac{ft}{sec}\right)^2 - (145.2m)(s) = \frac{1}{2}(0)^2$$

$$(145.2)(s) = \frac{1}{2}(117.3)^2$$

$s = 47.4$ ft

Answer: 47.4 ft

Example 3.17

Barrel barriers (Figure 3.15) are very often used on roads and highways to reduce crash fatalities. For a barrel barrier, the relation between the barrel opposing force and deflection of the barrel barrier is $F = 5,000,000x$ where x is the penetration of the barrel barrier in ft and F is in lb. A car with a weight of 6,000 lb is traveling at a speed of 100 ft/sec just before it hits the barrier. When the car hits the barrel barrier, the driver takes his feet off the accelerator (which is common). The coefficient of kinetic friction between the road surface and the car tire is 0.20. Determine the car's maximum penetration of the barrier.

FIGURE 3.15 Barrel Barrier

SOLUTION

- The speed of the car just before it crashes into the barrier is $v_1 = 100$ ft/sec.
- The maximum penetration occurs when the car is brought to a stop, i.e., $v_2 = 0$.
- The barrel barrier works to resist car penetration ($F.x$).
- The friction force between the road surface and the car also works to resist the car penetration ($F_f.x$).
- When the car hits the barrel barrier, the driver takes his feet off the accelerator (applied force is zero).

Mass, $m = \dfrac{6,000\,\text{lb}}{32.2\,\dfrac{\text{ft}}{\text{sec}^2}} = 186.34\,\text{slug}$

Initial kinetic energy, $T_1 = \dfrac{1}{2}mv_1^2 = \dfrac{1}{2}(186.34\,\text{slug})\left(100\,\dfrac{\text{ft}}{\text{sec}}\right)^2 = 931,700\,\text{lb.ft}$

Final kinetic energy, $T_2 = \dfrac{1}{2}mv_2^2 = \dfrac{1}{2}(186.34\,\text{slug})(0)^2 = 0$

Work done on the system, $U_{1\to2} = \sum Fs$

$$= (-F_f)(x\,\text{ft}) - \int_0^x (5,000,000x)\,dx$$

$$= (-\mu mg)(x\,\text{ft}) - \left[\frac{5,000,000x^2}{2} + C\right]_0^x$$

$$= -(0.2)(mg)(x) - \left[\left(\frac{5,000,000x^2}{2} + C\right) - (0 + C)\right]$$

$$= -(0.2)(6,000\,\text{lb})x - 2,500,000x^2$$

$$= -1,200x - 2,500,000x^2$$

We know: $T_1 + U_{1\to2} = T_2$

$$\frac{1}{2}mv_1^2 + U_{1\to2} = \frac{1}{2}mv_2^2$$

$$931,700 - 1,200x - 2,500,000x^2 = 0$$

$$9,317 - 12x - 25,000x^2 = 0$$

$$25,000x^2 + 12x - 9,317 = 0$$

Solving, $x = 0.61$ ft or 7.32 in.

Answer: 7.32 in.

Example 3.18

A 100-kg box, initially at rest, is subjected to the forces shown in Figure 3.16. Determine the distance it slides in order to attain a speed of 5 m/sec. The coefficient of kinetic friction between the box and the surface is 0.2.

FIGURE 3.16   The 100-Kg Box for Example 3.18

SOLUTION

Suppose, after sliding a distance, s, it will attain a velocity of 5 m/sec.

The box is initially at rest. That means the initial kinetic energy is zero. Then, some work is done on the box, and some work is lost by friction. After this, the box gains kinetic energy.

Let us find out the frictional force.

Apply $\sum F_y = 0$ ($\uparrow$ +ve) in the free-body diagram shown in Figure 3.17.

$R - mg - 350\ \text{N}\ \sin40° + 600\ \text{N}\ \sin60° = 0$

$R = (100\ \text{kg})\ 9.81\ \text{m/sec}^2 + (350\ \text{N})\ \sin40° - (600\ \text{N})\ \sin60°$

$R = 686.4\ \text{N}$

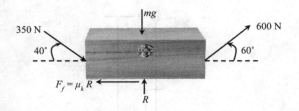

FIGURE 3.17   Analysis of the 100-Kg Box for Example 3.18

$F^f = \mu_k R$
$F^f = 0.2 \times 686.4 \text{ N}$
$F^f = 137.3 \text{ N}$

Initial kinetic energy, $T_1 = \dfrac{1}{2}mv_1^2 = \dfrac{1}{2}m(0)^2 = 0$

Final kinetic energy, $T_2 = \dfrac{1}{2}mv_2^2 = \dfrac{1}{2}(100\text{kg})\left(5\dfrac{\text{m}}{\text{sec}}\right)^2 = 1{,}250\text{ J}$

Work done on the system, $U_{1\rightarrow 2} = \sum Fs = [F_1 + F_2 - F_f]s$

$$= \left[(600\text{N})\cos 60 + (350\text{N})\cos 40 - 137.3\right]s$$
$$= 430.8s$$

Apply the work-energy principle:

$$T_1 + U_{1\rightarrow 2} = T_2$$

$0 + 430.8s = 1{,}250\text{ J}$
$s = 2.90 \text{ m}$
Answer: 2.9 m

3.6.2.4   Potential Energy

The energy acquired due to the application of a force on a body or energy acquired by the body due to the change of position of its particles is called the potential energy of the body. Suppose a piece of brick is kept above a roof, or water is pumped into a tank placed on a roof. In both cases, they acquired some energy. This type of energy is called potential energy. Potential energy is measured by the available work when the body returns to its initial normal position. Let an object of mass, m, be raised above the surface of the Earth to a height, dh, against gravity, as shown in Figure 3.18.

Work done due to this movement:

$$d\vec{W} = \vec{F}.d\vec{h}$$

$$dW = F.dh$$

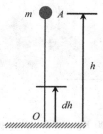

FIGURE 3.18   An Object Raised above the Surface against Gravity

The total work done in raising the body to a height, h, at position A is the sum of all differential work done as:

$$V = \int dW = \int_0^h F\,dh$$

$$= \int_0^h mg\,dh$$

$$= mg\int_0^h dh$$

$$= mg\,[h]_0^h$$

$$= mgh - mg(0)$$

$$= mgh$$

The SI unit for potential energy is J (N.m), and the US customary unit is lb.ft.

Example 3.19

If a horizontal force of 20 N is applied to a body with a mass of 1 kg, it moves 3 m along the upward inclined plane shown in Figure 3.19. Determine the work done on the body.

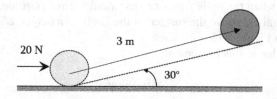

FIGURE 3.19   Movement of the Body for Example 3.19

SOLUTION

The body moved on an inclined path with both horizontal and vertical components of movement (Figure 3.20).

Horizontal displacement, $s = (3 \text{ m}) \cos30°$
Vertical displacement, $h = (3 \text{ m}) \sin30°$
Work done, $W = F.s + mgh$
$= (20 \text{ N}) (3 \cos30°) + (1 \text{ kg}) (9.81 \text{ m/sec}^2) (3 \sin30°)$
$= 66.7 \text{ J}$

Answer: 66.7 J

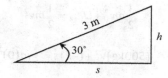

FIGURE 3.20

Example 3.20

A 10,000-kg hammer falls from a height of 15 m on a 500-kg concrete pile, as shown in Figure 3.21. The pile enters 1.5 m into the ground after each hit of the hammer. Determine the initial velocity of the pile when the hammer hits it.

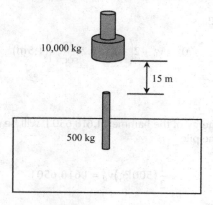

FIGURE 3.21   Hitting a Concrete Pile with a Hammer

SOLUTION

When the hammer hits the pile, the potential energy of the hammer converts into kinetic energy.
Hammer drop height = 15 m
Pile penetration = 1.5 m

Total fall, $h = 15$ m + 1.5 m = 16.5 m
Potential energy of the hammer = mgh = 10,000 kg (9.81 m/sec²) (16.5 m) = 1,618,650 J
Let F be the resisting force in N.
Work done by the resisting force, $W = F.s = F$ N(1.5 m) = 1.5F J
Work done by the resisting force = Work done by the hammer
1.5F = 1,618,650 J
Resisting force, F = 1,079,100 N

For pile:

$$T_1 + U_{1 \to 2} = T_2$$

$$\frac{1}{2}mv_1^2 + U_{1 \to 2} = \frac{1}{2}mv_2^2$$

$$\frac{1}{2}(500\,kg)v_1^2 + (-F)s = \frac{1}{2}m(0)^2$$

$$\frac{1}{2}(500\,kg)v_1^2 - (1,079,100\,N)(1.5m) = 0$$

v_1 = 80.46 m/sec

Alternative Solution 1

Deceleration of the pile, $a_o = F/m$ = 1,079,100 N / 500 kg = 2,158.2 m/sec²
Now, $v^2 = v_0^2 + 2a_o s$

$$0^2 = v_0^2 + 2\left(-2,158.2\frac{m}{sec^2}\right)(1.5m)$$

v_o = 80.46 m/sec

Alternative Solution 2
The potential energy of the hammer, 1,618,650 J, will be converted into the kinetic energy of the pile.

$$\frac{1}{2}(500\,kg)v_o^2 = 1,618,650\,J$$

v_o = 80.46 m/sec

Answer: 80.5 m/sec

Example 3.21

A 4-kg mass is allowed to fall freely from a height of 25 m by gravitational attraction. After 2 sec, determine the potential energy of the mass.

SOLUTION

Suppose the 4-kg mass travels a distance of y from the top and reaches a height of h above the ground after 2 sec (Figure 3.22).

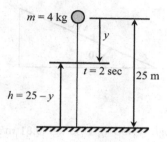

FIGURE 3.22

Potential energy, V after 2 sec = mgh

Travel after 2 sec, $y = v_0 t + \dfrac{1}{2} g t^2$

$$= 0 + \frac{1}{2}\left(9.81\frac{m}{sec^2}\right)(2\,sec)^2$$

$$= 19.6\,m$$

Height, h from the ground = 25 m – 19.6 m = 5.4 m

V after 2 sec = mgh = 4 kg (9.81 m/sec²)(5.4 m) = 212 J

Answer: 212 J

Example 3.22

A person with a mass of 150 kg steps along a stair of 4 m long with a load of 50 kg. If the inclination of the stair is 30° with the ground, as shown in Figure 3.23, determine the work done by the person.

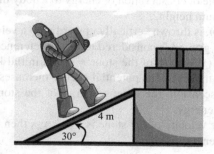

FIGURE 3.23 Stepping up along the Stair (Image by BrainPOP)

SOLUTION

Let the vertical distance traveled be h.
From Figure 3.23, $h = (4 \text{ m}) \sin 30°$ (Figure 3.24)

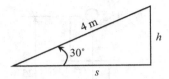

FIGURE 3.24

Work done, $W = mgh = (150 \text{ kg} + 50 \text{ kg}) (9.81 \text{ m/sec}^2) (4 \text{ m} \sin 30°) =$ 3,924 J

Answer: 3,924 J

3.6.2.5 Conservation of Energy

When two palms are rubbed together, they become warm. Here, mechanical energy is converted into heat energy. In another example, when a falling body strikes the ground and stops, mechanical energy is converted into heat energy and sound energy. Yet another example, due to friction between different parts of a machine, heat energy is created. Energy is converted from one form to another; however, it is never exhausted or destroyed. This is called the principle of conservation of energy. The law can be stated as 'Energy can neither be created nor destroyed but can only be converted from one form to another'. The total energy of the universe is constant.

In the case of a head-on collision between two cars, a similar energy conversion takes place. Before the collision, the cars only have kinetic energy; at the time of the collision, the two cars momentarily stop, and their kinetic energy becomes zero, i.e., kinetic energy changes to potential energy. After the collision, the cars move away from each other, and their potential energy changes back to kinetic energy again. This process can be explained with two springs. If two springs collide together and are released, they then move apart again. Now we will apply the principle of conservation of energy to a body that is thrown upward and reaches a maximum height.

A stone of mass, m, is thrown vertically upward with a velocity of v_o, as shown in Figure 3.25. If the ground is considered to be the reference level, then the initial potential energy ($V = mgh$) of the stone = 0 and initial kinetic energy (T) = $\frac{1}{2} mv_o^2$. As the stone moves up, its potential energy increases, and kinetic energy decreases. During ascension, the kinetic energy of the stone is gradually converted into potential energy.

At height, h, if the velocity of the stone is v ($v < v_o$), then the kinetic energy = $\frac{1}{2} mv^2$ and potential energy = mgh.

The total energy is = $V + T = mgh + \frac{1}{2} mv^2$.

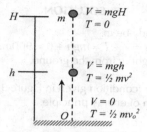

FIGURE 3.25 Stone Thrown Vertically Upward

When the stone reaches the highest point, H, its velocity becomes zero. Then, the kinetic energy (T) becomes zero, and potential energy (V) becomes mgH. At this point, all the kinetic energy converts into potential energy.

After reaching the maximum height, the stone starts to descend. Then, the opposite phenomenon occurs. The stone's potential energy starts decreasing as h decreases, and its kinetic energy starts increasing as v increases. When it reaches the ground (reference level), only kinetic energy exists, and potential energy becomes zero. Therefore, if there is no dissipative force like frictional force, the total energy of the stone at any point of its ascent or descent is the same. Mathematically, it can be shown as:

$$\text{Total energy at the top} = T + V = 0 + mgH = mgH$$

$$\text{Total energy at height}, h = T + V = \frac{1}{2}mv^2 + mgh$$

$$\text{Total energy at the ground} = T + V = \frac{1}{2}mv_o^2 + 0 = \frac{1}{2}mv_o^2$$

According to the conservation of energy principle, $mgH = \frac{1}{2}mv^2 + mgh = \frac{1}{2}mv_o^2$

If A and B are any two points in the travel path, then:

$$T_A + V_A = T_B + V_B$$

If any external work is done on the system from A to B ($U_{A \to B}$), then the above equation can be written as:

$$T_A + V_A + U_{A \to B} = T_B + V_B$$

Example 3.23

A body is falling from a height of 900 m due to gravitational attraction. Determine the height from the ground where its kinetic energy will be double that of its potential energy.

SOLUTION

Let the mass of the body be m.
Total energy at 900 m $= V + T = mgh + 0 = 900mg$
Also let, $T = 2V$ at height, y from the ground.
V at height, $y = mgy$
T at height, $y = 2\ mgy$ [condition given in Figure 3.26]
From the conservation of energy principle,

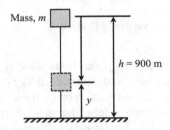

Mass, m

$h = 900$ m

y

FIGURE 3.26

Total Energy at y = Total Energy at the top
$mgy + 2\ mgy = 900mg$
$y = 300$ m

Answer: 300 m from the ground

Example 3.24

A body is allowed to fall freely from a height of 100 m. Determine the height where its potential energy will be three times the kinetic energy.

SOLUTION

Maximum potential energy $= mgh = 100mg$
At the height of y from the ground, $V = 3T$ [condition given]
V at the height of $y = mgy$ [Figure 3.27]

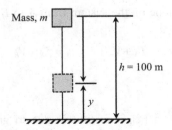

Mass, m

$h = 100$ m

y

FIGURE 3.27

T at the height of $y = \dfrac{V}{3} = \dfrac{mgy}{3}$

Total Energy at $y = V + T = mgy + \dfrac{mgy}{3}$

Total Energy at the top $= V + T = 100mg + 0 = 100mg$
From the conservation of energy principle,
Total Energy at y = Total Energy at the top

$$\frac{mgy}{3} + mgy = 100mg$$

$y = 75$ m

Answer: 75 m above the ground or 25 m down from the top

Example 3.25

A 35-kg boy starts from rest at A and travels down the ramp on a waterslide, as shown in Figure 3.28. If friction and air resistance are ignored, determine his speed when he reaches B, the edge of the ramp. Also, compute the horizontal distance s to where he strikes the water's surface at C if he makes the jump traveling horizontally at B. Ignore the boy's size.

 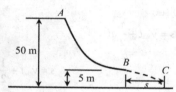

FIGURE 3.28 Path of the Boy

SOLUTION

From A to B:

Kinetic energy at A, $T_A = \dfrac{1}{2}mv_A^2 = \dfrac{1}{2}m(0)^2 = 0$

The potential energy at A, v_A

$=mgh_A = (35\,\text{kg})\left(9.81\dfrac{\text{m}}{\text{sec}^2}\right)(50\,\text{m}) = 17{,}167.5$ J

Kinetic energy at B, $T_B = \dfrac{1}{2}mv_B^2 = \dfrac{1}{2}(35\,\text{kg})v_B^2 = 17.5v_B^2$

The potential energy at B, $V_B = mgh_B = (35\,\text{kg})\left(9.81\dfrac{\text{m}}{\text{sec}^2}\right)(5\,\text{m}) = 1{,}716.75$ J

From the conservation of energy principle,
Total Energy at A = Total Energy at B

$$T_A + V_A = T_B + V_B$$

$$0 + 17{,}167.5 = 17.5v_B^2 + 1{,}716.75$$

$v_B = 29.71$ m/sec (horizontal)

From B to C:

The part BC of the travel path is a horizontal projectile. The boy starts
with a horizontal velocity of 29.71 m/sec from a height of 5 m and
gradually reaches the ground.

The horizontal distance, $s = v_{ox}t + \dfrac{1}{2}a_{ox}t^2$

Horizontal velocity, $v_{ox} = 29.71$ m/sec

There is no acceleration in the horizontal direction, i.e., $a_{ox} = 0$. The only
unknown to calculate s is time (t).

In the vertical direction, $y = v_{oy}t + \dfrac{1}{2}gt^2$

Vertical initial velocity, $v_{oy} = 0$

Vertical distance, $y = 5$ m (downward)

Vertical acceleration $= g$ (downward)

Therefore, $5\,\text{m} = (0)t + \dfrac{1}{2}\left(9.81\dfrac{\text{m}}{\text{sec}^2}\right)t^2$

$t = 1.0$ sec

Now, $s = v_{xo}t + \dfrac{1}{2}at^2$

$$s = v_{ox}t + \frac{1}{2}(0)t^2$$

$$s = \left(29.71\,\frac{\text{m}}{\text{sec}}\right)(1\text{sec}) + 0$$

$s = 29.71$ m

Answer: 29.7 m

Example 3.26

The pendulum, shown in Figure 3.29, is operating using a frictionless pivot.
The length of the massless rod is 5 m. If the pendulum is released from point A,
determine velocity (m/sec) at position B.

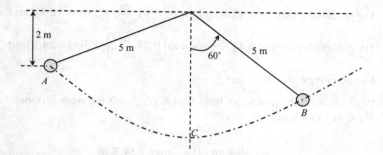

FIGURE 3.29 Travel Path of the Pendulum

SOLUTION

Total energy at A plus any work done on it equals the total energy at B, i.e.,

$$T_A + V_A + U_{A \to B} = T_B + V_B$$

Let the mass of the bob be m, and the equilibrium position (the lowest position, C) be the reference point. From the OBD triangle,

$$\cos 60° = \frac{h}{5\,\text{m}}$$

$h = 5$ m $\cos 60°$
Vertical height of A, $h_A = 5$ m $- 2$ m $= 3$ m
Vertical height of B, $h_B = 5$ m $- 5$ m $\cos 60° = 2.5$ m
The velocity of the bob at A, $v_A = 0$
The velocity of the bob at B, $v_B = ?$ (Figure 3.30)

Kinetic energy at A, $T_A = \frac{1}{2}mv_A^2 = \frac{1}{2}m(0)^2 = 0$

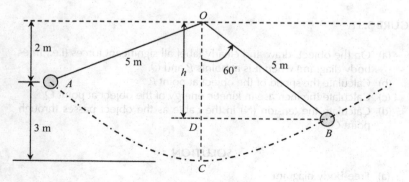

FIGURE 3.30 Analysis of the Travel Path of the Pendulum

The potential energy at A, $V_A = mgh_A = m\left(9.81\dfrac{m}{sec^2}\right)(3.0\,m) = 29.43m$ J

The potential energy at B, $V_B = mgh_B = m\left(9.81\dfrac{m}{sec^2}\right)(2.5\,m) = 24.53m$ J

Kinetic energy at B, $T_B = \dfrac{1}{2}mv_B^2$

Work done on the pendulum from A to B, $U_{A\to B} = 0$ [no work is done]

Now, $T_A + V_A + U_{A\to B} = T_B + V_B$

$$0 + 29.43m + 0 = \frac{1}{2}mv_B^2 + 24.53m$$

$$v_B = 3.13\frac{m}{sec}$$

Answer: 3.13 m/sec

Example 3.27

A 0.5-kg object rotates freely in a vertical circle at the end of a cable 2 m long, as shown in Figure 3.31. As the object passes through the point P at the top of the circular path, the tension in the cable is 20 N. Recall from Chapter 2, the force acting along the center of a circular curve on a mass, m rotating at velocity, v around a circular path of radius, R equals mv^2/R.

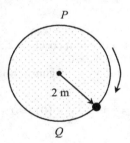

FIGURE 3.31

(a) On the object, draw and clearly label all significant forces (i.e., free-body diagram) when it is at points P and Q.
(b) Calculate the speed of the object at point P.
(c) Calculate the increase in kinetic energy of the object at point Q.
(d) Calculate the tension (N) in the cable as the object passes through point Q.

SOLUTION

(a) Free-body diagram
(b) Speed of the object at point P

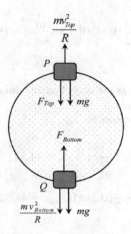

FIGURE 3.32 Cable Tensions at the Top and the Bottom of a Vertical Circle

From Figure 3.32, at P:
Cable tension, $F_{Top} = 20$ N
$\sum F_y = 0$

$$mg + F_{Top} = \frac{mv_{Top}^2}{R}$$

$$\frac{mv_{Top}^2}{R} - mg = 20\,\text{N}$$

$$\frac{(0.5\,\text{kg})v_{Top}^2}{2\,\text{m}} - (0.5\,\text{kg})\left(9.81\frac{\text{m}}{\text{sec}^2}\right) = 20\,\text{N}$$

$v_{Top} = 9.98$ m/sec (*Answer*)

(a) Kinetic energy of the object at point Q.

As the object moves from P to Q, it loses potential energy and gains kinetic energy. The gain in kinetic energy is equal to the loss in potential energy. Let the datum be the horizontal line passing through the lowest point, i.e., Q, of the circular path. Therefore, the potential energy at Q is zero.

Loss in potential energy at the bottom = $mg\Delta h$ [where, Δh = diameter of the circular path]
= (0.5 kg) (9.81 m/sec²) (4 m)
= 19.62 J

Kinetic energy at the bottom = $\frac{1}{2}mv_{Top}^2 + 19.62$ J

$$= \frac{1}{2}(0.5\,\text{kg})\left(9.98\,\frac{m}{\text{sec}}\right)^2 + 19.62\,\text{J}$$

=44.52 J (*Answers*)
(b)   Tension in the cable as the object passes through point *Q*.

Let the velocity at the bottom be v_{Bottom}

$$\frac{1}{2}mv_{Bottom}^2 = 44.52\,\text{J}$$

$$\frac{1}{2}(0.5\,\text{kg})v_{Bottom}^2 = 44.52\,\text{J}$$

$v_{Bottom} = 13.34$ m/sec
$\sum F_y = 0$

$$F_{Bottom} - mg - \frac{mv_{Bottom}^2}{R} = 0$$

Cable tension at the bottom, $F_{Bottom} = \dfrac{mv_{Bottom}^2}{R} + mg$

$$= \frac{(0.5\,\text{kg})\left(13.34\,\dfrac{m}{\text{sec}}\right)^2}{2\,m} + (0.5)\left(9.81\,\frac{m}{\text{sec}}\right)$$

= 49.40 N (*Answers*)

3.7   IMPULSIVE FORCE

Momentum is a measurement of mass in motion: How much mass is in how much motion. It is defined as the product of the mass and its velocity.

$$P = mv \;(\text{scalar form})$$

$$\vec{P} = m\vec{v}\,(\text{vector form})$$

As velocity is a vector quantity, momentum is also a vector quantity. The SI unit of momentum is kg.m/sec, and the US customary unit is slug.ft/sec. Impulse is a term that quantifies the overall effect of a force acting over time. It is conventionally given the symbol *J* and expressed as follows:

$$J = Ft \left(\text{scalar form}\right)$$

$$\vec{J} = \vec{F}t \left(\text{vector form}\right)$$

As force is a vector quantity, an impulse is also a vector quantity. The SI unit of momentum is N.sec, and the US customary unit is lb.sec. This type of force is active during a collision, explosion, sudden strike, etc. To hit a tennis ball, kick a soccer ball, hit a nail with a hammer, strike a string of a musical instrument, all are examples of a special type of force. This force is called impulsive force. It acts for a very short period, and during that time displacement of the body may be ignored. However, if the magnitude of the force is very large, there is a sudden change of velocity, and the momentum also changes. It is not possible to know or measure the impulsive force correctly. It is not needed. If the change or momentum can be measured, i.e., if the impulse of force is known, then the total result of the force can be known. For this reason, this type of force is called impulsive. This means that for a very brief period, if a large force acts on a body, then that force is called impulsive force. The product of the force and its duration of action is called the impulse of the force. For example, a particle of mass, m, is moving at a velocity of v_o. After being acted on by a force of F for t sec, the velocity will be v. Then, the impulse of a force can be expressed as:

$$J = Ft$$

$$J = mat$$

$$= m\left(\frac{v - v_o}{t}\right)t$$

$$= \frac{mv - mv_o}{t}t$$

$$= mv - mv_o = \text{change of momentum}$$

In vector form,

$$\vec{J} = m\left(\vec{v} - \vec{v}_o\right)$$

This is exactly equivalent to the change in momentum. This equivalency is known as the impulse-momentum theorem. Because of the impulse-momentum theorem, we can make a direct connection between how a force acts on an object over time and the motion of the object. It is also written as:

$$mv_1 + \sum F\left(\Delta t\right) = mv_2$$

This means the initial momentum of the particle plus the sum of all impulses applied equals the final momentum of the particle. One of the reasons why impulse is important is that, in the real world, forces are often not constant. Forces due to factors like people or engines tend to build up from zero over time and may vary. Working out the overall effect of all these forces directly would be quite difficult.

Example 3.28

A body of mass of 0.05 kg hits a vertical wall with a horizontal velocity of 0.2 m/sec and rebounds with a velocity of 0.1 m/sec. Determine the applied impulse.

SOLUTION

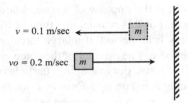

FIGURE 3.33

Mass, $m = 0.05$ kg (Figure 3.33)
Initial velocity, $v_0 = 0.2$ m/sec
Final velocity, $v = -0.1$ m/sec
Impulse, $J = m(v - v_0)$
$= 0.05$ kg $(-0.1 - 0.2)$
$= -0.015$ kg.m/sec

Answer: 0.015 kg.m/sec or N.sec

Example 3.29

The 3.0-Mg suburban is traveling with a constant speed of 75 mph by the precise use of the car accelerator. After taking the foot off the accelerator, the speed decreases from 75 mph to 20 mph in 5 sec. Determine the coefficient of kinetic friction between the tires and the road.

SOLUTION

Friction force always acts opposite to the motion. For a running car, friction always gives a resistive force. If a brake is applied, friction force helps the braking force to decrease the speed of the car.
  Mass of the car, $m = 3.0$ Mg $= 3,000$ kg
  Stopping time, $t = 5$ sec

Initial velocity, $v_1 = 75$ mph $= \dfrac{75\left(1.61\dfrac{km}{mile}\right)\left(1{,}000\dfrac{m}{km}\right)}{3{,}600\dfrac{sec}{h}} = 33.54\dfrac{m}{sec}$

Final velocity, $v_2 = 20$ mph $= \dfrac{20\left(1.61\dfrac{km}{mile}\right)\left(1{,}000\dfrac{m}{km}\right)}{3{,}600\dfrac{sec}{h}} = 8.94\dfrac{m}{sec}$

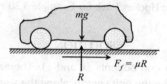

FIGURE 3.34 Free-Body Diagram of the Suburban

From Figure 3.34, $\displaystyle\sum F_y = 0$

$R - mg = 0$
$R = 3{,}000$ kg $(9.81$ m/sec$^2)$
$R = = 29{,}430$ N

Frictional force, $F_f = \mu\,(29{,}430)$ N
Now, $mv_1 + \displaystyle\sum F(\Delta t) = mv_2$

$$3{,}000 \text{ kg}\left(33.54\ \frac{m}{sec}\right) + (-29{,}430\ \mu)(5\sec) = 3{,}000 \text{ kg}\left(8.94\ \frac{m}{sec}\right)$$

$$\mu = 0.5$$

Answer: 0.5

Note that there is no brake force in this example. If there was a brake force, i.e., if a brake was applied, the brake force must be added in the $\displaystyle\sum F(\Delta t)$ component of the equation.

Example 3.30

A 0.2-kg hockey ball is traveling in northerly direction with a velocity of 10 m/sec. It is then struck by a hockey stick and given a velocity of 20 m/sec to the northeast, as shown in Figure 3.35. Determine the magnitude of the net impulse exerted by the hockey stick on the ball.

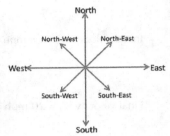

FIGURE 3.35   Striking the Hockey Puck for Example 3.30

SOLUTION

The direction has two components (x and y). Therefore, vector analysis is required. Let, $\vec{i}$ and $\vec{j}$ be the unit vectors along the x and y axes, respectively.

Initial velocity, $\vec{v}_o = 10\vec{j}$ (along the north)
Final velocity, $\vec{v} = 20\cos 45\,\vec{i} + 20\sin 45\,\vec{j}$ (along the north-east)
Impulse, $\quad\vec{J} = m(\vec{v} - \vec{v}_o) = 0.2\left[\left(20\cos 45\,\vec{i} + 20\sin 45\,\vec{j}\right) - 10\vec{j}\right]$

$\qquad = 2.83\vec{i} + 0.83\vec{j}$

Magnitude of the Impulse, $J = \sqrt{(2.83)^2 + (0.83)^2} = 2.95\ \dfrac{\text{kg.m}}{\text{sec}}$

Answer: 2.95 kg.m/sec

3.8   APPLICATION OF IMPULSIVE FORCE

When a rocket moves upward while in flight, you will see that a smoke-like white cloud comes out through the end of its nozzles. Can you say why such smoke can be seen? Due to the burning of fuel, gas is formed at very high pressure. We often see this coil of gas from Earth. This gas comes out through the small opening at the rear side of the rocket with tremendous velocity. Due to this, an equally tremendous reaction force is produced, which pushes the rocket upward. Although the mass of the exhaust gas is small, the momentum of the gas is very high due to its velocity. According to the conservation principle of momentum, the rocket also acquires equal but oppositely directed momentum and thus rises upward with a high velocity. Typically, liquid hydrogen is used to fuel a rocket. For ignition, the liquid hydrogen together with oxygen are allowed to enter the combustion chamber. Due to burning of the fuel, high-pressure gas is produced that comes out at a very high speed through the opening at the rear of the rocket.

Let us consider that a rocket is in motion in space where we can ignore air resistance and the influence of gravity. Due to the emission of gas from the rocket, a force or a thrust against the motion of the gas is generated, which pushes the rocket ahead at a very high speed.

Let the applied thrust = F
Mass of the rocket = M
Mass of the ejected gas at the time of $\Delta t = \Delta m$
The escape velocity of gas = v
Change of momentum of the gas at the time interval of $\Delta t = \Delta m(v - 0) =$
$(\Delta m)v$
According to the conservation principle of momentum, change in momentum at the time of Δt = applied impulse force on the rocket

$$(\Delta m)v = F(\Delta t)$$

$$F = \left(\frac{\Delta m}{\Delta t}\right)v$$

If the instantaneous acceleration of the rocket is a, then $F = Ma$, thus:

$$a = \frac{F}{M} = \frac{1}{M}\left(\frac{\Delta m}{\Delta t}\right)v$$

If gravitational acceleration is considered,

$$F = \left(\frac{\Delta m}{\Delta t}\right)v - Mg$$

$$Ma = \left(\frac{\Delta m}{\Delta t}\right)v - Mg$$

$$a = \frac{F}{M} = \frac{1}{M}\left(\frac{\Delta m}{\Delta t}\right)v - g$$

where $\dfrac{\Delta m}{\Delta t}$ is the rate of fuel consumption.

From the equation, it can be seen that if the mass of the rocket decreases, the velocity increases. To increase the acceleration of the rocket, the rate of the ejection of the gas has to be increased. When the relative velocity of the gas increases, acceleration also increases.

Example 3.31

A rocket consumes 0.07 kg fuel per sec. If the velocity of gas ejected by the rocket is 100 km/sec, determine the force acting on the rocket ignoring the gravitational force.

SOLUTION

Rate of fuel use, $\dfrac{\Delta m}{\Delta t} = 0.07 \dfrac{kg}{sec}$

Force acting on the rocket, $F = \left(\dfrac{\Delta m}{\Delta t}\right)v = \left(0.07 \dfrac{kg}{sec}\right)\left(100,000 \dfrac{m}{sec}\right) =$

7,000 N
Answer: 7,000 N

Example 3.32

A rocket, while moving upward, loses 1/50th part of its mass during the first 2 sec. If the velocity of the ejected gas is 2,500 m/sec, calculate the acceleration of the rocket considering gravitational action as well.

SOLUTION

Loss of mass, $\Delta m = \dfrac{M}{50}$
Time, $\Delta t = 2$ sec

$$F = \left(\dfrac{\Delta m}{\Delta t}\right)v - Mg$$

$$Ma = \left(\dfrac{\Delta m}{\Delta t}\right)v - Mg$$

$$a = \dfrac{1}{M}\left(\dfrac{\Delta m}{\Delta t}\right)v - g$$

$$= \dfrac{1}{M}\left(\dfrac{M}{50 \times 2\,sec}\right)2,500\dfrac{m}{sec} - 9.81\dfrac{m}{sec^2}$$

$$= 15.2\dfrac{m}{sec^2}$$

Answer: 15.2 m/sec²

3.9 CONSERVATION OF LINEAR MOMENTUM

If no external force is acting on bodies or if a net applied external force is zero, the total linear momentum of those bodies remains constant, or there will not be any change of momentum. For example, when a bullet is fired from a gun, it goes ahead with tremendous velocity. The initial momentum was zero before firing as both the gun and the bullet were at rest. However, after firing the gun, the bullet acquires momentum. The gun acquires an equal but oppositely directed momentum. Mathematically, the principle of conservation of linear momentum can be verified. Suppose two particles of masses m_1 and m_2 are moving along the same direction in a straight line with velocities of u_1 and u_2, respectively, with $u_1 > u_2$. At one time, the first particle strikes the second particle from behind, as shown in Figure 3.36. Afterward, the two particles continue to move in the same direction in a straight line with velocities, v_1 and v_2, respectively.

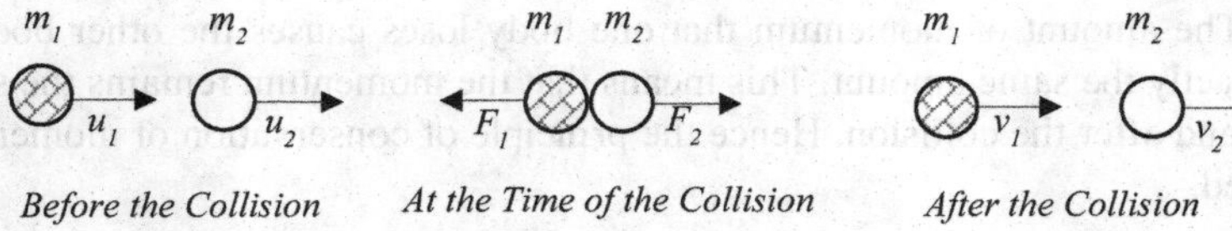

Before the Collision   *At the Time of the Collision*   *After the Collision*

FIGURE 3.36   Movement and Striking of the Particles

Let the time of collision of action and reaction be t.

Summation of the initial momentum of the two particles $= m_1u_1 + m_2u_2$

Summation of the final momentum of the two particles $= m_1v_1 + m_2v_2$

From the laws of conservation of momentum, it can be proved that,

$$m_1u_1 + m_2u_2 = m_1v_1 + m_2v_2$$

Proof:

The rate of change of momentum of the first particle is

$$= \frac{m_1v_1 - m_1u_1}{t}$$

$$= m_1\left(\frac{v_1 - u_1}{t}\right)$$

$$= \text{Reaction Force}$$

$$= F_1$$

The rate of change of momentum of the second particle is

$$= \frac{m_2v_2 - m_2u_2}{t}$$

$$= \text{Action Force}$$

$$= F_2$$

But the rate of change of momentum of the two bodies is equal and opposite, i.e., $F_2 = -F_1$

$$\frac{m_2v_2 - m_2u_2}{t} = -\frac{m_1v_1 - m_1u_1}{t}$$

After rearranging, it can be shown that:

$$m_1 u_1 + m_2 u_2 = m_1 v_1 + m_2 v_2$$

The sum of initial momentum of the two bodies = The sum of final momentum of the two bodies.

Due to the action and reaction forces, no change of total momentum has taken place. The amount of momentum that one body loses causes the other body to gain exactly the same amount. This means that the momentum remains the same before and after the collision. Hence the principle of conservation of momentum is proved.

3.10   IMPACT

An impact is a high force or shock applied over a short period when two or more bodies collide. Such a force or acceleration usually has a greater effect than a lower force applied over a proportionally longer period. The effect depends, critically, on the relative velocity of the bodies to one another. For example, a nail is pounded with a series of impacts, each by a single hammer blow. These high-velocity impacts overcome the static friction between the nail and the substrate. Although on a much larger scale, a pile driver achieves the same result by driving a pile into the soil. This method is commonly used during civil construction projects to make the foundations for buildings and bridges. Road traffic accidents usually involve impact loading, such as when a car hits a traffic bollard, water hydrant, or tree, the damage is being localized to the impact zone. If the force that the objects exert on each other is parallel to the path of motion and directed toward the center of gravity of each object, the objects undergo a direct central impact. Otherwise, it is called oblique central impact.

During an impact, momentum is conserved while particular energy may or may not be conserved. For direct central impact with no external forces:

$$m_1 v_1 + m_2 v_2 = m_1 v_1' + m_2 v_2'$$

where:
m_1, m_2 = the masses of the two bodies
v_1, v_2 = the velocities of the bodies just before impact
v_1', v_2' = the velocities of the bodies just after impact

For impacts, the relative velocity expression is:

$$e = \frac{\left(v_2'\right) - \left(v_1'\right)}{\left(v_1\right) - \left(v_2\right)}$$

where:
e = coefficient of restitution

The coefficient of restitution (*e*) is the ratio of the final relative velocity to the initial relative velocity between two objects after they collide. It is a positive real number, and it typically ranges between 0 and 1, where 1 would be a perfectly elastic collision, and 0 would be a perfectly inelastic collision. The value of *e* is such that:

e = 0: This is a perfectly inelastic collision. The objects do not move apart after the collision, but instead, they stay together. Kinetic energy is converted to heat or work while the objects are being deformed.

0 < *e* < 1: This is a real-world inelastic collision in which some kinetic energy is dissipated.

e = 1: This is a perfectly elastic collision, in which no kinetic energy is dissipated, and the objects rebound from one another with the same relative speed with which they approached.

Example 3.33

A force of 16 N acts on a body of 4-kg mass for 4 sec. Determine the change of velocity and the impulse of force.

SOLUTION

Given,
Force, $F = 16$ N
Mass, $m = 4$ kg
Time, $t = 4$ sec
Impulse, $J = Ft = m(v - v_o)$
Then, $(v - v_o) = \dfrac{Ft}{m} = \dfrac{16\,\text{N}(4\,\text{sec})}{4\,\text{kg}} = 16\dfrac{m}{sec}$
Impulse, $J = Ft = (16\,\text{N})(4\,\text{sec}) = 64\,\text{N.sec}$

Answers: 16 m/sec, 64 N.sec

Example 3.34

A 1,500 kg car is traveling at 10 m/sec. Another 2,000 kg car is traveling at 15 m/sec toward the first car, as shown in Figure 3.37. After the collision, determine the combined velocity.

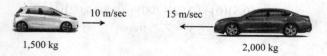

FIGURE 3.37 Traveling Information of the Cars

SOLUTION

Given,

Mass of the first car, $m_1 = 1,500$ kg
Velocity of the first car, $v_1 = 10$ m/sec (e.g., positive)
Mass of the second car, $m_2 = 2,000$ kg
Velocity of the second car, $v_2 = -15$ m/sec (negative as opposite to that of the first car)
The combined velocity, $v' = ?$

Known: $m_1 v_1 + m_2 v_2 = (m_1 + m_2) v'$

$$1,500\,\text{kg}\left(10\frac{m}{sec}\right) + 2,000\,\text{kg}\left(-15\frac{m}{sec}\right) = \left(1,500\,\text{kg} + 2,000\,\text{kg}\right)v'$$

$$15,000 - 30,000 = 3,500 v'$$

Therefore, $v' = -4.3$ m/sec
The negative sign implies that the velocity after the collision is in the direction of travel of the second car.

Answer: 4.3 m/sec in the direction of travel of the second car.

Example 3.35

A bird having a mass of 500 g is sitting in a tree. A bullet of mass 20 g struck the bird at a horizontal velocity of 200 m/sec. Determine the horizontal velocity of the bird if the bullet remains inside the bird.

SOLUTION

Given,

Mass of the bird, $m_1 = 500$ g $= 0.5$ kg
Velocity of the bird, $v_1 = 0$
Mass of the bullet, $m_2 = 20$ g $= 0.02$ kg
Velocity of the bullet, $v_2 = 200$ m/sec
The combined velocity, $v' = ?$

Known: $m_1 v_1 + m_2 v_2 = (m_1 + m_2) v'$

$$(0.5)(0) + (0.02)(200) = (0.5 + 0.02)v'$$
$$v' = 7.7 \quad \text{m/sec}$$

Answer: 7.7 m/sec

Example 3.36

One 1-kg ball, A, and another 2-kg ball, B, collide, as shown in Figure 3.38. If the coefficient of restitution of the balls is 0.75, for each ball just after the collision, determine the following:

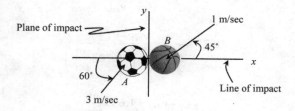

FIGURE 3.38   Colliding of Two Balls for Example 3.36

(a) x-components of the final velocity
(b) y-components of the final velocity
(c)   The final velocities and their directions

SOLUTION

Let subscript-1 represent the initial condition, and subscript-2 represent the final condition. As this is an oblique impact, we need to resolve the velocity components along the x-axis as the x-axis is the line of impact.

$$v_{Ax1} = 3\cos 60 = 1.5\ \frac{m}{sec}$$

$$v_{Ay1} = 3\sin 60 = 2.6\ \frac{m}{sec}$$

$$v_{Bx1} = -1\cos 45 = -0.707\ \frac{m}{sec}$$

$$v_{By1} = -1\sin 45 = -0.707\ \frac{m}{sec}$$

(a) Applying the conservation of momentum principle along the x-direction,

$$m_A v_{Ax1} + m_B v_{Bx1} = m_A v_{Ax2} + m_B v_{Bx2}$$

$$(1\text{kg})\left(1.5\frac{m}{sec}\right) + (2\text{kg})\left(-0.707\frac{m}{sec}\right) = (1\text{kg})v_{Ax2} + (2\text{kg})v_{Bx2}$$

$$v_{Ax2} + 2v_{Bx2} = 0.086 \frac{m}{sec} \tag{1}$$

Now,

$$e = \frac{v_{Bx2} - v_{Ax2}}{v_{Ax1} - v_{Bx1}}$$

$$0.75 = \frac{v_{Bx2} - v_{Ax2}}{1.5 \frac{m}{sec} - \left(-0.707 \frac{m}{sec}\right)}$$

$$v_{Bx2} - v_{Ax2} = 1.655 \frac{m}{sec} \tag{2}$$

Solving Equations (1) and (2),

$$v_{Bx2} = 0.58 \frac{m}{sec} (\rightarrow)$$

$$v_{Ax2} = -1.07 \frac{m}{sec} = 1.07 \frac{m}{sec} (\leftarrow)$$

(b) In the y-direction, there is no external impulse. Therefore, no velocity change will occur along the y-direction.

$$v_{Ay2} = v_{Ay1} = 2.6 \frac{m}{sec} (-)$$

$$v_{By2} = v_{By1} = -0.707 \frac{m}{sec} = 0.707 \frac{m}{sec} (\downarrow)$$

(c) The combined velocities:

$$v_{B2} = \sqrt{(v_{Bx2})^2 + (v_{By2})^2} = \sqrt{\left(0.58 \frac{m}{sec}\right)^2 + \left(-0.707 \frac{m}{sec}\right)^2} = 0.91 \frac{m}{sec}$$

$$\theta = \tan^{-1} \left|\frac{v_{By2}}{v_{Bx2}}\right| = \tan^{-1} \left|\frac{0.707}{0.58}\right| = 50.6°$$

This angle is clockwise with respect to the positive x-axis, as $v_{Bx2} \rightarrow$ and $v_{By2} \downarrow$.

$$v_{A2} = \sqrt{(v_{Ax2})^2 + (v_{Ay2})^2} = \sqrt{\left(-1.07 \frac{m}{sec}\right)^2 + \left(2.6 \frac{m}{sec}\right)^2} = 2.81 \frac{m}{sec}$$

$$\theta = \tan^{-1}\left|\frac{v_{Ay2}}{v_{Ax2}}\right| = \tan^{-1}\left|\frac{2.6}{1.07}\right| = 67.6°$$

This angle is clockwise with respect to the negative x-axis, as $v_{Ax2} \leftarrow$ and $v_{Ay2} \uparrow$.

Answers:

(a) $v_{Ax2} = 1.07\ \dfrac{m}{sec}(\leftarrow),\ v_{Bx2} = 0.58\dfrac{m}{sec}(\rightarrow)$

(b) $v_{Ay2} = v_{Ay1} = 2.6\dfrac{m}{sec}(\uparrow),\ v_{By2} = 0.707\ \dfrac{m}{sec}(\downarrow)$ (Figure 3.39)

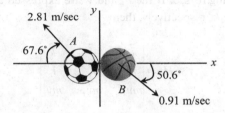

FIGURE 3.39   Answers of Example 3.36(c)

3.11   ANGULAR MOMENTUM AND ANGULAR IMPULSE

Angular Momentum: Angular momentum of a particle is defined as the moment of a particle's linear momentum about an axis. Thus, sometimes, it is referred to as the moment of momentum. For example, consider a particle with mass m is moving in a curvilinear path in the x-y plane. We are interested in finding the angular moment about point O due to the linear moment of this particle. Consider that r is the perpendicular distance from point O to the line of action of the motion at point A, as shown in Figure 3.40.

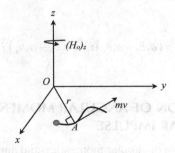

FIGURE 3.40   Movement in the Curvilinear Path

At the instant shown, if the velocity of the particle is v, the linear momentum of the particle will be mv. Thus, the angular momentum about point O can be expressed as:

$$\left(H_o\right)_z = \left(mv\right)r$$

The direction of angular momentum will be perpendicular to the x-y plane at point O, i.e., along the positive z-axis.

In vector form, this angular momentum can be expressed as:

$$\vec{H}_o = \vec{r} \times m\vec{v}$$

The SI unit of angular momentum is N.m.sec or kg.m^2/sec and the US customary unit is lb.ft.sec or slug.ft^2/sec. If the $\vec{r}$ and $\vec{v}$ are expressed as $\vec{r} = r_x\vec{i} + r_y\vec{j} + r_z\vec{k}$ and $\vec{v} = v_x\vec{i} + v_y\vec{j} + v_z\vec{k}$, respectively, then:

$$\vec{H}_o = \begin{vmatrix} \vec{i} & \vec{j} & \vec{k} \\ r_x & r_y & r_z \\ mv_x & mv_y & mv_z \end{vmatrix}$$

Angular Impulse: If a force $\vec{F}$ is applied to a body from time t_1 to t_2, and $\vec{r}$ is a position vector that originates from point O to the line of action of force at any point, then the angular impulse of the above system can be expressed in a vector form as:

$$\int_{t_1}^{t_2} M_o dt = \int_{t_1}^{t_2} \left(\vec{r} \times \vec{F}\right) dt$$

Knowledge of the cross-product of vectors is required to apply the above equation. Let us assume two vectors: $\vec{A} = a_x\vec{i} + a_y\vec{j} + a_z\vec{k}$ and $\vec{B} = b_x\vec{i} + b_y\vec{j} + b_z\vec{k}$. The cross-product of these two vectors is a vector whose direction is perpendicular to both vectors and can be determined as:

$$\begin{vmatrix} \vec{i} & \vec{j} & \vec{k} \\ a_x & a_y & a_z \\ b_x & b_y & b_z \end{vmatrix} = \left(a_y b_z - a_z b_y\right)\vec{i} + \left(a_z b_x - a_x b_z\right)\vec{j} + \left(a_x b_y - a_y b_x\right)\vec{k}$$

3.12   CONSERVATION OF ANGULAR MOMENTUM AND ANGULAR IMPULSE

The conservation principle of angular momentum and angular impulse states that the initial angular momentum of a particle plus any angular impulse done on the particle equals the final angular momentum of the particle. In vector form, it can be written as:

$$\left(\vec{H}_1\right)_o + \sum \int_{t_1}^{t_2} \vec{M}_o dt = \left(\vec{H}_2\right)_o$$

In scalar form, it can be written as:

$$\left(H_1\right)_{x,y,z} + \sum \int_{t_1}^{t_2} M_{x,y,z} dt = \left(H_2\right)_{x,y,z}$$

Example 3.37

The supporting rod of the system in Figure 3.41 is subjected to a couple moment of $M = 10t^2$ N.m where t is in sec. Determine the speed of the car after 5 sec, if the car starts from rest. Given that the total mass of the car and the rider is 150 kg, and the lever arm is 3 m.

FIGURE 3.41 Engine System for Example 3.37

SOLUTION

Mass, $m = 150$ kg
Initial velocity, $v_1 = 0$
Radius, $r = 3$ m
Time, $t = 5$ sec
Final velocity, $v = ?$
Initial momentum, $H_1 = mv_1 r_1 = 0$
Final momentum, $H_2 = mv_2 r_2 = \left(150\,\text{kg}\right)v\left(3\,\text{m}\right) = 450v$
$\displaystyle\int_{t_1}^{t_2} M_z dt$ by the applied moment $= \displaystyle\int_0^{5\,\text{sec}} \left(10t^2\right)dt = \left[\frac{10t^3}{3}\right]_0^5 = \frac{10\left(5\right)^3}{3} - 0$

$\qquad = 416.67\,\text{kg.m}^2/\text{sec}$

From the principle of angular impulse and momentum:

$$\left(H_1\right)_o + \sum \int_{t_1}^{t_2} M_o\, dt = \left(H_2\right)_o$$

$$0 + \int_0^{5\,\text{sec}} \left(10t^2\right) dt = 450v$$

416.67 = 450v
v = 0.93 m/sec

Answer: 0.93 m/sec

Example 3.38

The supporting rod of the system in Figure 3.42 is subjected to a couple moment of $M = 12t$ N.m and assume each engine supplies a force of $F = 6t$ N to the car where t is in sec. Determine the speed of the car after 5 sec, if the car starts from rest. Given that the total mass of the car and the rider is 200 kg, and the length of the lever arm is 3 m.

FIGURE 3.42   Engine System for Example 3.38

SOLUTION

Mass, $m = 200$ kg
Initial velocity, $v_1 = 0$
Radius, $r = 3$ m

Time, $t = 5$ sec
Final velocity, $v = ?$
Initial momentum, $H_1 = mv_1 r_1 = 0$
Final momentum, $H_2 = mv_2 r_2 = (200 \, kg)(v)(3 \, m) = 600v$

$$\int_{t_1}^{t_2} M_z dt \text{ by the applied moment} = \int_0^{5\,sec} (12t)dt = \left[\frac{12}{2}t^2\right]_0^5 = \frac{12}{2}(5)^2 - 0 = 150$$

kg.m²/sec

$$\int_{t_1}^{t_2} M_z dt \text{ by the applied force} = \int_0^{5\,sec} F(r)dt = \int_0^{5\,sec} 6t(3)dt$$

$$= \int_0^{5\,sec} (18t)dt = \left[9t^2\right]_0^5 = 9(5)^2 - 0 = 225 \text{ kg.m}^2/sec$$

Now, principle of angular impulse and momentum:

$$(H_1)_o + \sum \int_{t_1}^{t_2} M_o dt = (H_2)_o$$

$0 + 150 + 225 = 600v$
$v = 0.625$ m/sec

Answer: 0.625 m/sec

Example 3.39

An amusement park ride consists of a 200-kg car and passenger and has a radius of 4 m. The car travels at 3 m/sec along a circular path, and then the cable is pulled toward the center at 0.1 m/sec. Determine the speed of the car after 8 sec. Also, determine the work done to pull on the cable.

SOLUTION

Mass, $m = 200$ kg
Initial velocity, $v_1 = 3$ m/sec
Initial radius, $r_1 = 4$ m
Time, $t = 8$ sec

Radius after 8 sec, $r_2 = 4m - \left(0.1\frac{m}{sec}\right)(8\,sec) = 3.2m$

Final velocity after 8 sec, $v = ?$
Consider the tangential velocity of the car with the passenger after 8 sec
as v_{2t}
Initial angular momentum,

$$H_1 = mv_1 r_1 = 200 \text{ kg}\left(3\frac{m}{sec}\right)(4m) = 2,400\frac{kg.m^2}{sec}$$

Final angular momentum, $H_2 = mv_2r_2 = (200\,\text{kg})(v_{2t})(3.2\,\text{m}) = 640v_{2t}$

Applying Conservation of Angular Momentum:

$$640v_{2t} = 2,400$$

$$v_{2t} = 3.75\ \frac{\text{m}}{\text{sec}}$$

This is tangential velocity. The radial velocity is 0.1 m/sec toward the center of the system.
Final velocity,

$$v_2 = \sqrt{(v_{2t})^2 + (v_{2r})^2} = \sqrt{\left(3.75\frac{\text{m}}{\text{sec}}\right)^2 + \left(0.1\frac{\text{m}}{\text{sec}}\right)^2} = 3.751\frac{\text{m}}{\text{sec}}$$

Applying Principle of Work and Energy:

$$T_1 + \sum U_{1\to2} = T_2$$

$$\frac{1}{2}mv_1^2 + \sum U_{1\to2} = \frac{1}{2}mv_2^2$$

$$\frac{1}{2}(200\,\text{kg})(3\text{ m/sec})^2 + \sum U_{1\to2} = \frac{1}{2}(200\,\text{kg})(3.751\text{ m/sec})^2$$

$$\sum U_{1\to2} = 507\text{ J}$$

Answers: 3.751 m/sec, 507 J

FUNDAMENTALS OF ENGINEERING (FE) EXAM STYLE QUESTIONS

FE Problem 3.1

For a rocket ship in deep space, far away from any other objects, to move in a straight line with a constant speed ignoring the air resistance or gravity, it must exert a net force that is:

A. proportional to its mass
B. proportional to its weight
C. proportional to its velocity
D. zero

E. proportional to its displacement

FE Problem 3.2

A loaded truck collides with a car, causing a large amount of damage to the car. Which of the following is true about the collision?

    A. the force on the truck is greater than the force on the car
    B. the force on the car is greater than the force on the truck
    C. the force on the truck is the same in magnitude as the force on the car
    D. during the collision, the truck has a greater displacement than the car
    E. during the collision, the truck has a greater acceleration than the car

FE Problem 3.3

A car is parked on the side of a street. The net force on the car can be expressed as which of the following?

    A. non-zero vector pointing up
    B. non-zero vector pointing down
    C. non-zero vector pointing left
    D. non-zero vector pointing right
    E. it is zero

FE Problem 3.4

A roller coaster ride at an amusement park lifts a car at point A to a height of 90 m above the lowest point on the track, as shown in Figure 3.43. The car starts from rest at point A, rolls with negligible friction down the incline, and follows the track. Given that point B is on the ground, point C is 40 m above the ground, and point D is 25 m above the ground. The fastest speed of the coaster would be at _____, and the value would be _____ respectively:

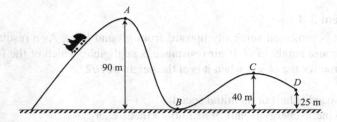

FIGURE 3.43   A Roller Coaster Ride at an Amusement park

    A. Point B, 42 m/sec
    B. Point A, 24 m/sec
    C. Point D, 48 m/sec
    D. not enough information to solve the problem

FE Problem 3.5

A football is kicked off at the ground at a distance of 50 yards downfield. Ignoring air resistance, which of the following statements would be INCORRECT when the football reaches the highest point?

A. all of the ball's original kinetic energy has been changed into potential energy
B. the ball's horizontal velocity is the same as when it left the kicker's foot
C. the ball will have been in the air one-half of its total flight time
D. the vertical component of the velocity is equal to zero

FE Problem 3.6

A rock is dropped from the top of a tall tower. Half a second later, another rock, twice as big as the first one, is dropped. Identify the most accurate sentence below. Ignore air resistance.

A. the distance between the rocks increases while both are falling
B. the acceleration is greater for the rock with the bigger mass
C. they strike the ground more than half a second apart
D. they strike the ground with the same kinetic energy

FE Problem 3.7

A rock is lifted for a certain amount of time by a force F that is greater in magnitude than the rock's weight, W. The change in kinetic energy of the rock during this time is equal to the:

A. work done by the net force $(F-W)$
B. work done by F alone
C. work done by W alone
D. difference in the potential energy of the rock before and after this time

FE Problem 3.8

An object is projected vertically upward from ground level. As a result, it rises to a maximum height of H. If air resistance is negligible, which of the following must be true for the object when it is at the height of $H/2$?

A. its speed is half of its initial speed
B. its kinetic energy is half of its initial kinetic energy
C. its potential energy is half of its initial potential energy
D. its total mechanical energy is half of its initial value

FE Problem 3.9

A 2-kg block is released from rest at the top of a curved incline in the shape of a quarter circle with radius R, as shown in Figure 3.44. The block then slides onto

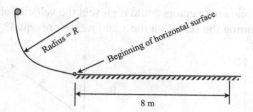

FIGURE 3.44 An Irregular Surface

a horizontal plane where it finally comes to rest 8 m from the beginning of the horizontal surface. The curved incline is frictionless, but there is an 8-N force of friction on the block while it slides horizontally. The radius in meters of the inclined surface is most nearly:

A. 1.6
B. 2.6
C. 3.2
D. 3.8

FE Problem 3.10

A 10-kg block is pushed along a rough horizontal surface by a constant horizontal force, F, as shown in Figure 3.45. At time $t = 0$, the velocity v of the block is 6.0 m/sec in the same direction as the force. The coefficient of friction is 0.20.

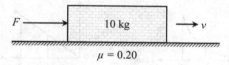

FIGURE 3.45 A Block Being Pushed by a Force

The force, F, in N necessary to keep the velocity constant is most nearly:

A. 9.81
B. 19.62
C. 23.65
D. 29.43

FE Problem 3.11

An object of mass, m, is initially at rest and free to move without friction in any direction in the x-y plane. A constant net force of magnitude F directed in the $+x$ direction acts on the object for 1 sec. Immediately thereafter, a constant net force of the same magnitude, F, directed in the $+y$ direction, acts on the object for 1 sec. After this, no forces act on the object.

Which of the following vectors could represent the velocity of the object at the end of 3 sec, assuming the scales on the x and y axes are equal?

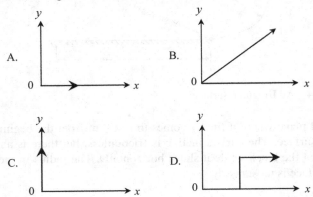

FE Problem 3.12

An object of mass, m, is initially at rest and free to move without friction in any direction in the x-y plane. A constant net force of magnitude F directed in the $+x$ direction acts on the object for 1 sec. Immediately thereafter, a constant net force of the same magnitude F directed in the $+y$ direction acts on the object for 1 sec. After this, no forces act on the object. Which of the following graphs best represents the kinetic energy ($T = \frac{1}{2} mv^2$) of the object as a function of time?

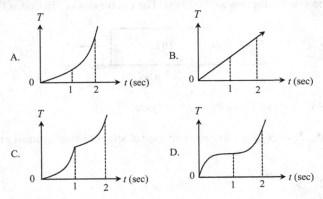

FE Problem 3.13

A 5-kg particle is moving with a velocity of $\vec{v} = 3\vec{i} + 2\vec{j} - 5\vec{k}$ m/sec. If the particle's position vector is given as $\vec{r} = 10\vec{i} - 2\vec{j} + 5\vec{k}$ m, determine the angular momentum (kg.m/sec^2) of the particle at its origin.

A. $100\vec{i} + 325\vec{j} + 130\vec{k}$

B. $-325\vec{j}+130\vec{k}$

C. $325\vec{j}+130\vec{k}$

D. $-100\vec{i}+325\vec{j}+130\vec{k}$

FE Problem 3.14

A 30-g bullet is fired with a speed of 500 m/sec into a wall. If the deceleration of the bullet is constant and it penetrates 12 cm into the wall, the force (N) on the bullet while it is stopping is most nearly:

A. 3,750
B. 12,250
C. 23,250
D. 31,250

FE Problem 3.15

A 30-g bullet is fired with a speed of 500 m/sec into a wall. The deceleration of the bullet is constant and it penetrates 12 cm into the wall. The force (N) on the bullet, while it is stopping, is approximately 31,250 N. The time (sec) required for the bullet to stop is most nearly:

A. 0.02
B. 0.0048
C. 0.00048
D. 0.48

FE Problem 3.16

Two objects are moving on a frictionless horizontal surface. Object P initially moves at 10 m/sec, as shown in Figure 3.46a. It then collides elastically with another identical object Q, which is initially at rest. After the collision, object P moves at 6 m/sec along a path at 50° to its original direction.

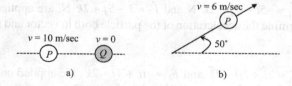

a)   b)

FIGURE 3.46 Two Objects in a Collision for FE Problem 3.16

Which of the following diagrams best represents the motion of object Q after the collision?

A. ------Q------

$v = 0$

B.
40°

Q

$v = 8$ m/sec

C.

Q $v = 8$ m/sec

40°

D.

50°

Q

$v = 8$ m/sec

FE Problem 3.17

When two elastic bodies collide with each other:

A. the two bodies will momentarily come to rest after the collision
B. the two bodies tend to compress and deform at the surface of contact
C. the two bodies begin to regain their original shape
D. all of the above

PRACTICE PROBLEMS

Section I: Force and Newton's Laws

Problem 3.1

A force of 50 N acts on a body at rest with a mass of 10 kg. If the force does not act after 4 seconds, what distance will the body travel in 8 seconds from the start?

Problem 3.2

A body with a mass of 3 kg moves with uniform acceleration and travels a distance of 0.18 m and 0.30 m in the 5th and 8th sec, respectively. Determine the force that is acting on the body.

Problem 3.3

Two forces, $\vec{F_1} = -5\vec{i} - 2\vec{j} - 2\vec{k}$ N and $\vec{F_2} = \vec{i} - 5\vec{j} + 2\vec{k}$ N, are applied on a 750-g particle. Determine the acceleration of the particle both in vector and scalar forms.

Problem 3.4

Two forces, $\vec{F_1} = 2\vec{i} + 2t\vec{j} - 7\vec{k}$ and $\vec{F_2} = -t\vec{i} + 3\vec{j} - 2\vec{k}$, are applied on a 10-kg particle at rest, where the forces are in N and t is in sec. Determine the following:

(a) the acceleration of the particle after 2 sec
(b) the velocity of the particle after 2 sec
(c) the distance traveled in 2 sec

Problem 3.5

A block with a mass of 5 kg was initially at rest at point A on an inclined plane. A force F = 40 N was applied to the block as shown below. What will be the velocity of the block at point B? Given that $\theta = 30°$, s = 2.5 m, and the coefficient of kinetic friction is 0.15 (Figure 3.47).

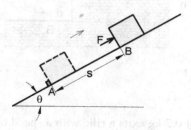

FIGURE 3.47 Problem 3.5

Problem 3.6

A block with a mass of 4 kg was placed on an inclined plane at point A, and it started sliding due to gravity, as shown below. How far will it travel along the plane after 2 sec, i.e., s = ? Given that $\theta = 30°$ and the coefficient of kinetic friction is 0.2 (Figure 3.48).

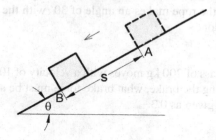

FIGURE 3.48 Problem 3.6

Problem 3.7

If the 10-kg block on the inclined surface starts sliding from rest at point A and hits the horizontal surface at point B, determine the distance d. Given that $\theta = 30°$, s = 3.5 m, h = 2 m, and the coefficient of kinetic friction between the inclined surface and the block is 0.15 (Figure 3.49).

Problem 3.8

A bullet with a mass of 0.01 kg was fired with a velocity of 400 m/sec from a gun with a mass of 5 kg. Calculate the recoil velocity of the gun.

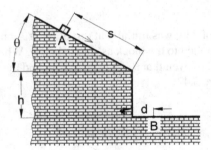

FIGURE 3.49   Problem 3.7

Problem 3.9

A bullet with a mass of 0.3 kg exits a rifle with a speed of 30 m/sec. If the rifle tends to recoil with a speed of 0.6 m/sec, calculate the mass of the rifle.

Problem 3.10

A bullet is fired at a speed of 80 cm/sec from a gun with a mass of 20 kg. If the mass of the bullet is 40 g, then find the velocity of the gun.

Section II: Work and Energy

Problem 3.11

A piece of ice is dragged 3 m on a horizontal plane by a rope. If the tension in the rope is 10 N and the rope makes an angle of 30° with the surface plane, then determine the work done.

Problem 3.12

A motor car with a mass of 200 kg moves with a velocity of 108 km/h. To stop the car in 20 m by applying the brake, what brake force must be applied? The kinetic friction coefficient is given as 0.3.

Problem 3.13

A truck with a mass of 5 tons is moving at a speed of 36 km/h. To stop the truck in 4 m, how much force needs to be applied?

Problem 3.14

When a force acts on a body with a mass of 40 kg at rest, it starts moving, and after 2 sec, its velocity becomes 15 m/sec. Find the magnitude of the force applied on the body and its kinetic energy after 4 sec.

Problem 3.15

Determine the magnitude of the force that needs to be applied on a body with a mass of 200 kg so that its velocity increases from $\left(4\vec{i} - 5\vec{j} + 3\vec{k}\right)$ m/sec to $\left(8\vec{i} + 3\vec{j} - 5\vec{k}\right)$ m/sec after traveling 200 m.

Problem 3.16
A truck with a mass of 900 kg moves with a velocity of 60 km/h. The truck stops at a distance of 50 m by applying the brake. If the frictional force of the ground is 200 N, calculate the magnitude of the force due to the applied brake.

Problem 3.17
A boy weighing 250 N performs 2,000 J of work to climb vertically to the top of a ladder. Find the length of the ladder.

Problem 3.18
A trolley weighing 40 kg was moving on a smooth road with a kinetic energy of 180 J. An object with a mass of 20 kg was dropped on the trolley. Determine the final kinetic energy.

Problem 3.19
A hammer with a mass of 2 kg hits vertically on a nail placed on a wall. At what velocity does the nail need to be struck so that the nail enters 0.025 m through the wall while overcoming a resistive force of 640 N?

Problem 3.20
1 N of force is applied against the motion of a body having a kinetic energy of 1 J. How far will it move before stopping?

Problem 3.21
A bullet can penetrate a piece of wood of a given thickness. How many times of the original velocity would be needed to penetrate 16 of these wood pieces together?

Problem 3.22
At what velocity does a bullet with a mass of 0.05 kg eject in producing 365 J of kinetic energy from a gun with a mass of 3.6 kg?

Problem 3.23
If the velocity of a body with a mass of 10-lbm is $\left(7\vec{i} - 6\vec{j} + 5\vec{k}\right)$ ft/sec, determine its kinetic energy.

Problem 3.24
A body is allowed to fall freely from a height of 300 m by gravitational attraction. Determine the height where its kinetic energy becomes equal to half of its potential energy.

Problem 3.25
A body with a mass of 200 g drops from a height of 10 m. Determine the kinetic energy of the body just before touching the ground.

Problem 3.26

A body with a mass of 500 g falls from the top of a ship onto the water 10 m below. Calculate the following:

(a) the initial potential energy of the body
(b) the maximum kinetic energy
(c) the maximum velocity during the fall
(d) kinetic energy at 3 m above the water surface
(e) the potential energy at 3 m above the water surface

Problem 3.27

A body with a mass of 2 kg falls from a height of 5 m. How much work is done on the body by gravitational force, and how much potential energy is lost from the body?

Problem 3.28

From what height would a body with a mass of 2 kg need to fall due to gravitational force so that it acquires a kinetic energy of 2,400 J just before it touches the ground?

Problem 3.29

A 6-kg hammer falls from a height of 5 m onto a 400-gm pin, the pin enters 10 cm into the ground. Determine the initial kinetic energy of the pin when the hammer hits it. Also, how long does the pin take to penetrate the ground?

Problem 3.30

A 1-kg stone is thrown vertically upward with a velocity of 15 m/sec. Calculate the kinetic energy at the maximum height. What is its kinetic energy just before touching the ground when it returns?

Problem 3.31

A 100-kg crate is being pulled by a heavy truck with a force having a constant direction and a magnitude of $F = 1,500$ N, as shown in Figure 3.50. When $s = 10$ m, the truck is moving ahead with a speed of 2 m/sec. Determine its speed when $s = 25$ m. The coefficient of kinetic friction between the crate and the ground is 0.25 (Figure 3.50).

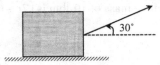

FIGURE 3.50   Crate Being Pulled for Problem 3.31

Section III: Impulse, Momentum, and Impact

Problem 3.32

An iron sphere with a mass of 2 kg, while moving horizontally with a velocity of 5 m/sec, hits a wall perpendicularly and rebounds with a velocity of 3 m/sec. Calculate the impulse.

Problem 3.33

A player kicked a football with a mass of 0.6 kg while in motion with a velocity of 25 m/sec. Consequently, the ball acquired a velocity of 40 m/sec in the same direction. Determine the applied impulse of a force by the foot of the player.

Problem 3.34

A cricket player stops a cricket ball with a mass of 0.2 kg coming at a speed of 20 m/sec in 0.1 sec by catching it. Determine the average force applied by the player.

Problem 3.35

A rocket burns 7.4 kg of fuel per sec. If the velocity of the emitted gas from the rocket is 2,500 m/sec, what is the magnitude of the force that acts on the rocket?

Problem 3.36

A body with a mass of 5 kg while moving at a velocity of 10 m/sec combines with another body with a mass of 2 kg moving in the same direction with a velocity of 3 m/sec. After combining into a single body, what will its velocity be?

Problem 3.37

Two bodies with masses of 40 kg and 60 kg moving opposite to each other at velocities of 10 m/sec and 2 m/sec, respectively, collide. After the collision, at what velocity will the combined body move?

Problem 3.38

A body with a mass of 5 kg is moving toward the north with a velocity of 4 m/sec. Another body with a mass of 3 kg is moving toward the south with a velocity of 2 m/sec. In a given time, the two bodies collide and become a single body. At what velocity and at what direction will the combined body move?

Problem 3.39

A 0.5-lb hockey ball is traveling to the west at a velocity of 12 ft/sec when it is struck by a hockey stick with a velocity of 20 ft/sec to the southeast, as shown below. Determine the magnitude of the net impulse exerted by the hockey stick on the ball (Figure 3.51).

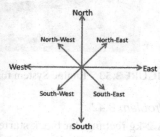

FIGURE 3.51   Striking the Hockey Ball for Problem 3.39

Problem 3.40

A 1-kg ball, *A*, and a 2-kg ball, *B*, collide, as shown in the following figure. If the coefficient of restitution of the balls is 0.75, for each ball just after the collision, determine the final velocities (Figure 3.52).

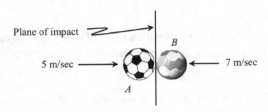

FIGURE 3.52   Colliding of Two Balls for Problem 3.40

Problem 3.41

The supporting rod of the system in Figure 3.53 is subjected to a couple moment of $M = 12t^2$ N.m, and the engine supplies a force of $F = 6t+3$ N to each car where *t* is in seconds. The total mass of the car and the rider is 200 kg, and the lever arm is 3 m long.

  (a) determine the speed of the car after 5 sec, if the car starts from rest
  (b) if the car has an initial velocity of 5 m/sec, how much brake force needs
       to be applied to stop the car in 10 sec without stopping the engine?
  (c) if the car has an initial velocity of 5 m/sec, how much brake force needs
       to be applied to stop the car in 10 sec in addition to stopping the engine?
       (Figure 3.53)

FIGURE 3.53   Engine System for Problem 3.41

Problem 3.42

A 2-kg rectangular block started rotating about the vertical axis when a constant force *F* was applied on the block, as shown in Figure 3.54. If the velocity of the

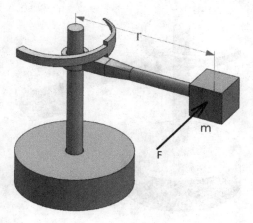

FIGURE 3.54 Problem 3.42

block is 4.2 m/s after 3 sec, determine the magnitude of the applied force. Given that $r = 0.75$ m.

Problem 3.43

As shown in Figure 3.54, a 2-kg rectangular block started rotating about the vertical axis when a force $F = 4t - 5$ N was applied on the block. How long will it take to reach a velocity of 2.8 m/s for the block? Given that $r = 0.75$ m.

Problem 3.44

A system of four cylindrical blocks started rotating about the vertical axis in a counterclockwise direction due to a couple moment $M = 54$ N.m that was applied to it as shown in Figure 3.55. The mass of each cylindrical block is 2.25 kg, and

FIGURE 3.55

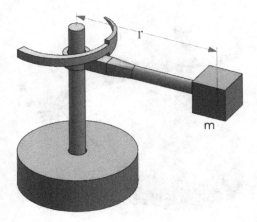

FIGURE 3.56

the center of mass for each block from the axis of rotation is 2.5 m. What will be the angular velocity of the system after 3 sec?

Problem 3.45

A 2-kg rectangular block rotates in a counterclockwise direction with a velocity of 3.5 m/s about the vertical axis, as shown in Figure 3.56. If a clockwise couple moment of 3.5 N.m is applied to the system, how long will it take to stop the rotation? Given that $r = 1.25$ m.

4 Kinematics of a Rigid Body

4.1 GENERAL

In Chapter 2, the kinematics of particles were discussed, which is particularly important for mechanical design and analysis. This chapter will present how the equations of motion can be applied to the kinematics of a rigid body in a plane. Note that the kinematics of a rigid body in space (three dimensions) is not within of the scope of this book. A body can be considered rigid when the distance between any two particles always remains the same during motion. A rigid body may undergo three major types of motions:

(a) Translation
(b) Rotation about a fixed axis
(c) General plane motion.

These three motions of a rigid body are discussed below.

4.2 TRANSLATION

Translation of a rigid body occurs when all of the line segments of a body remain parallel to its initial position during motion. A simple example of translation is the movement of your desk laterally by a distance. As shown in Figure 4.1, X-Y is the fixed coordinate system, and x-y is the translating coordinate system.

Position. The locations of point A and point B are defined from the fixed X-Y reference frame by using the position vectors $\vec{r}_A$ and $\vec{r}_B$. The translating x-y coordinate system is attached to the body and has its origin located at B.

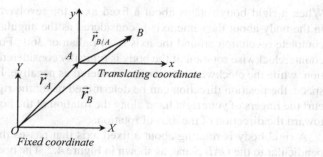

FIGURE 4.1 Fixed and Translating Coordinate Systems

DOI: 10.1201/9781003283959-4

The position vector, $\vec{r}_B$, can be expressed as:

$$\vec{r}_B = \vec{r}_A + \vec{r}_{B/A}$$

Where $\vec{r}_{B/A}$ is the position of B with respect to A

Velocity. The velocity can be obtained by differentiating the position vectors as follows:

$$\frac{d}{dt}\left(\vec{r}_B\right) = \frac{d}{dt}\left(\vec{r}_A\right) + \frac{d}{dt}\left(\vec{r}_{B/A}\right)$$

$$\vec{v}_B = \vec{v}_A + 0$$

$\frac{d}{dt}\left(\vec{r}_{B/A}\right) = 0$ as the position of B with respect to A does not change

The above velocity equation can be expressed as:

$$\vec{v}_B = \vec{v}_A$$

where $\vec{v}_B$ and $\vec{v}_A$ are the velocities measured from the *X-Y* (Fixed Coordinate) System. This equation indicates that all points in a rigid body during rectilinear or curvilinear translation move at the same velocity.

Acceleration. Like velocity, the acceleration can be presented as:

$$\vec{a}_A = \vec{a}_B$$

This equation indicates that all points in a rigid body during rectilinear or curvilinear translation move at the same acceleration.

4.3   ROTATION ABOUT AN ARBITRARY FIXED AXIS

4.3.1   ANGULAR DISPLACEMENT

When a rigid body rotates about a fixed axis, the revolved angle by any point in the body about the same axis is considered as the angular displacement. One complete revolution around the axis is 2π radian or $360°$. For a planar motion, a counterclockwise rotation of the body is typically considered as the positive rotation, while the clockwise rotation is considered as negative. In three-dimensional space, the position direction can be determined using the right-hand rule. If you curl the fingers of your right hand along the rotation of the body, the thumb points toward the direction of the axis of rotation.

A rigid body is rotating about a fixed axis that passes through O and is perpendicular to the *OAB* plane, as shown in Figure 4.2. The body rotates such that a

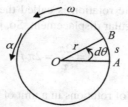

FIGURE 4.2 The Rotation of a Body about a Fixed Axis

line on the plane initially at OA is rotated counterclockwise to OB with an angular distance of $d\theta$ during time t.

If the angular displacement ($d\theta$) is expressed in radians, then its relationship with the associated arc s is linear and can be written as:

$$d\theta = \frac{s}{r}$$

4.3.2 Angular Velocity

If, during the exceedingly small time interval, Δt, the angular displacement of the particle is $\Delta \theta$, then the angular velocity for that time interval is:

$$\omega = \frac{\Delta\theta}{\Delta t}$$

In order to know the instantaneous velocity at a particular instant, the time interval is made smaller and smaller. If the time interval decreases toward zero, the average angular velocity during that time interval is equal to the instantaneous angular velocity. Therefore, the rate change of instantaneous angular displacement is called the instantaneous angular velocity, ω. This means:

$$\omega = \underset{\Delta t \to 0}{Lim} \frac{\Delta\theta}{\Delta t} = \frac{d\theta}{dt}$$

Usually, the angular velocity means the instantaneous velocity. If the angular velocity is constant, then that circular motion is called uniform circular motion. In the case of uniform circular motion, the angular velocity can be written as:

$$\omega = \frac{\theta}{t}$$

$$\theta = \omega t$$

This equation is similar to the uniform linear motion of $s = vt$. The angular velocity is expressed in radians/sec. It is also called rotation or revolutions per minute (rpm).

Time taken for a complete rotation is called the time period; one complete rotation means 2π radian angular displacement. So, if the time period is τ, then:

$$\omega = \frac{2\pi}{\tau} = 2\pi f$$

where, $f = \dfrac{1}{\tau}$ is the number of rotations in a unit of time or frequency. Its unit is sec^{-1} or Hz.

Angular velocity (ω) indicates how fast an object rotates or revolves with respect to another point. For example, how quickly the angular position or orientation of an object changes with time. Angular velocity (ω) is measured as angular displacement per unit of time, e.g., radians per second. If $d\theta$ is small, the length AB can be determined as $ds = rd\theta$. Then the linear velocity can be determined as $v = \dfrac{ds}{dt} = \dfrac{rd\theta}{dt} = r\left(\dfrac{d\theta}{dt}\right) = r\omega = \omega r$.

Therefore, $v = \omega r$

The equation says that the linear velocity is the product of angular velocity and the radius of the circular path.

Example 4.1

A particle revolves 120 times per minute in a circular path with a radius of 1.5 m. Determine the:

(a) Linear velocity
(b) Time period
(c) Angular velocity.

SOLUTION

Radius of rotation, $r = 1.5$ m

(a) Linear velocity, $v = \omega r = (2\pi n)r = 2\pi\left(\dfrac{120}{60\,\text{sec}}\right)(1.5\,\text{m}) = 18.85\dfrac{\text{m}}{\text{sec}}$

(b) Time period, $\tau = \dfrac{1}{f}$

n = the number of rotations in the unit time = 120 times / 60 sec = 2 sec^{-1}

Time period, $\tau = \dfrac{1}{f} = \dfrac{1}{2\,\text{sec}^{-1}} = 0.5\,\text{sec}$

(c) Angular velocity, $\omega = 2\pi f = 2\pi\left(\dfrac{120}{60\,\text{sec}}\right) = 12.57\,\text{rad/sec}$

Answers: 18.85 m/sec, 2 per sec, 12.57 rad/sec

4.3.3 ANGULAR ACCELERATION

The angular acceleration is defined as:

$$\alpha = \frac{d\omega}{dt} = \frac{d^2\theta}{dt^2}$$

Angular acceleration, (α), is expressed as the rate of change of angular velocity (ω). The unit of angular acceleration is presented in terms of angle per time squared. The linear acceleration (a) can be determined as follows:

$$\frac{dv}{dt} = \frac{d}{dt}(\omega r)$$

$$a = \omega \frac{d}{dt}(r) + r \frac{d}{dt}(\omega)$$

$$a = \omega(0) + r\alpha$$

$$a = \alpha r$$

The equation says linear acceleration is the product of angular acceleration and the radius of the circular path. From the relationship of $\omega = \dfrac{d\theta}{dt}$ and $\alpha = \dfrac{d\omega}{dt}$ the following relationship can be derived:

$$\alpha d\theta = \omega d\omega$$

This linear acceleration is along the tangential direction and is often called tangential acceleration (a_t), as shown in Figure 4.3. Therefore,

$$a_t = \alpha r$$

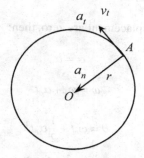

FIGURE 4.3 Travel along a Circular Path

The linear acceleration along the normal direction can be determined as follows:

$$a_n = \frac{v^2}{r} = \frac{(\omega r)^2}{r} = \omega^2 r$$

The magnitude of the linear acceleration can be determined as follows:

$$a = \sqrt{(a_t)^2 + (a_n)^2}$$

4.3.4   FOR CONSTANT ACCELERATION

Applying Newton's law of motion for constant angular acceleration, the equations for angular velocity and displacement are:

$$\alpha = \alpha_o$$

$$\omega = \omega_o + \alpha_o (t - t_o)$$

$$\theta = \theta_o + \omega_o (t - t_o) + \frac{1}{2} \alpha_o (t - t_o)^2$$

$$\omega^2 = \omega_o^2 + 2\alpha_o (\theta - \theta_o)$$

where:
θ = angular displacement (rad)
θ_o = angular displacement at time t_o (rad)
ω = angular velocity (rad/sec)
ω_o = angular velocity at time t_o (rad/sec)
α_o = constant angular acceleration (rad/sec^2)
t = time (sec)
t_o = some initial time (sec)

If the initial time and displacement are zero, then:

$$\alpha = \alpha_o$$

$$\omega = \omega_o + \alpha_o t$$

$$\theta = \omega_o t + \frac{1}{2} \alpha_o t^2$$

$$\omega^2 = \omega_o^2 + 2\alpha_o \theta$$

4.3.5 Non-constant Acceleration

When non-constant acceleration, $\alpha\,(t)$ is considered, the equations for the velocity and displacement may be obtained from:

$$v(t) = v_{t_o} + \int_{t_o}^{t} \alpha(t)\,dt$$

$$s(t) = s_{t_o} + \int_{t_o}^{t} v(t)\,dt$$

For variable angular acceleration:

$$\omega(t) = \omega_{t_o} + \int_{t_o}^{t} \alpha(t)\,dt$$

$$\theta(t) = \theta_{t_o} + \int_{t_o}^{t} \omega(t)\,dt$$

where t is the variable of integration.

Example 4.2

An electric motor is rotating at 6,000 rpm when the switch is suddenly turned off. The fan decelerates at a constant rate and comes to rest after 3 min. Calculate the angular deceleration of the electric fan.

SOLUTION

Given,

Initial angular velocity, $\omega_o = 6,000\,\text{rpm} = 6,000\,\dfrac{\text{rev}}{\text{min}}\left(2\pi\,\dfrac{\text{rad}}{\text{rev}}\right)$

$= 12,000\pi\,\dfrac{\text{rad}}{\text{min}}$

Elapsed time, $t = 3$ min
Final angular velocity, $\omega = 0$ (as it stops)
Angular deceleration, $\alpha_o = ?$
Known, $\omega = \omega_0 + \alpha_o t$
Or, $0 = 12,000\pi + \alpha_o(3\,\text{min})$

Therefore, $\alpha_o = -4,000\pi\,\dfrac{\text{rad}}{\text{min}^2}$

Answer: $\alpha_o = -4,000\pi\,\dfrac{\text{rad}}{\text{min}^2}$

Example 4.3

The angular position of a car traveling around a curve is described by $\theta(t) = 15t + 2t^2 - 6t^3 + 5$, where t is in sec and θ is in radians. Calculate the angular acceleration of the car at the beginning of the curve.

SOLUTION

Given,

Angular position at any time, $\theta(t) = 15t + 2t^2 - 6t^3 + 5$
Angular displacement,

$$\omega = \dot{\theta} = \frac{d\theta}{dt} = \frac{d}{dt}\left(15t + 2t^2 - 6t^3 + 5\right) = 15 + 4t - 18t^2$$

Angular acceleration, $\alpha = \dot{\omega} = \ddot{\theta}$

Therefore, $\alpha = \dfrac{d\omega}{dt} = \dfrac{d}{dt}\left(15 + 4t - 18t^2\right) = 4 - 36t$

At, $t = 0$ sec, $\alpha = 4$ rad/sec²
Answer: 4 rad/sec²

Example 4.4

An electric motor is rotating at 6,000 rpm when the switch is suddenly turned off. The fan decelerates at an angular deceleration of $4,000\,\pi\ \dfrac{rad}{min^2}$. Calculate how many revolutions the fan will undergo before stopping.

SOLUTION

Given,

Initial angular velocity,

$$\omega_o = 6,000\,\text{rpm} = 6,000\,\frac{rev}{min}\left(2\pi\,\frac{rad}{rev.}\right) = 12,000\pi\,\frac{rad}{min}$$

Final angular velocity, $\omega = 0$ (as it stops)

Angular deceleration, $\alpha = -4,000\pi\,\dfrac{rad}{min^2}$

Known, $\omega^2 = \omega_o^2 + 2(-\alpha)\theta$

$$0^2 = \left(12,000\pi\,\frac{rad}{min}\right)^2 + 2\left(-4,000\pi\,\frac{rad}{min^2}\right)(\theta)$$

$\theta = 18,000\pi$ rad
$\theta = 18,000\pi/2\pi$ rev [1 rev = 2π]
$\theta = 9,000$ rev

Answer: 9,000 revolutions

Note: 1 revolution = 2π radian or 360°

Example 4.5

The angular acceleration of a 0.4-ft diameter circular disk can be represented as $\alpha = 5t^2 - 6t + 2$ rad/sec^2, where t is in sec. The original angular velocity of the disk was 6 rad/sec. Determine the following after 5 sec of rotation:

(a) The linear velocity
(b) The linear acceleration

SOLUTION

Given,

Radius, r = 0.4 ft / 2 = 0.2 ft
Angular acceleration, $\alpha = 5t^2 - 6t + 2$ rad/sec^2

Both the linear velocity and linear acceleration require angular velocity. Therefore, let us determine the angular velocity first.

$$\alpha = \frac{d\omega}{dt} = 5t^2 - 6t + 2$$

$$\int d\omega = \int \left(5t^2 - 6t + 2\right)dt$$

$$\int_{6}^{\omega} d\omega = \int_{0}^{5\ sec} \left(5t^2 - 6t + 2\right)dt$$

$$[\omega]_{6}^{\omega} = \left[\frac{5t^3}{3} - 3t^2 + 2t\right]_{0}^{5\ sec}$$

$$[\omega - 6] = \left[\frac{5(5\sec)^3}{3} - 3(5\sec)^2 + 2(5\sec)\right] - \left[\frac{5(0)^3}{3} - 3(0)^2 + 2(0)\right]$$

$$\omega = 143.3 \ \frac{rad}{sec}$$

(a) Linear velocity, $v = \omega r = \left(143.3\dfrac{rad}{sec}\right)(0.2\,ft) = 28.67\dfrac{ft}{sec}$

(b) Linear acceleration

$$\alpha\,|_{t=5\,sec} = \left[5t^2 - 6t + 2\right]_{t=5\,sec} = 5(5\,sec)^2 - 6(5\,sec) + 2 = 97\dfrac{rad}{sec^2}$$

Tangential acceleration, $a_t = \alpha r = \left(97\dfrac{rad}{sec^2}\right)(0.2\,ft) = 19.4\dfrac{ft}{sec^2}$

Normal acceleration, $a_n = \omega^2 r = \left(143.3\dfrac{rad}{sec}\right)^2(0.2\,ft) = 4{,}107\dfrac{ft}{sec^2}$

Linear acceleration,

$$a = \sqrt{(a_t)^2 + (a_n)^2} = \sqrt{\left(19.4\dfrac{ft}{sec^2}\right)^2 + \left(4{,}107\dfrac{ft}{sec^2}\right)^2} \approx 4{,}107\dfrac{ft}{sec^2}$$

Answers: 28.67 fps, 4,107 ft/sec²

Example 4.6

The power of a car engine is transmitted using the belt-and-pulley arrangement, as shown in Figure 4.4. If the engine turns pulley *A* at 50 rad/sec, determine the angular velocities of pulley *B* and pulley *D*. The hub at *C* is rigidly connected to *B* and turns with it. The diameters of pulley *A*, *B*, *C*, and *D* are 0.5 ft, 0.7 ft, 0.2 ft, and 0.4 ft, respectively.

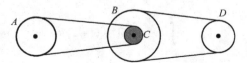

FIGURE 4.4   Belt-and-Pulley Arrangement for Example 4.6

SOLUTION

The peripheries of pulley *A* and *C* are connected with a belt. This linear speed is transferred to the periphery of pulley *C*.

Therefore, $v_A = v_C$

$$v_A = v_C$$

$$\omega_A r_A = \omega_C r_C$$

$$\omega_C = \dfrac{\omega_A r_A}{r_C} = \dfrac{50\dfrac{rad}{sec}(0.25\,ft)}{0.1\,ft} = 125\dfrac{rad}{sec}$$

Again, $\omega_C = \omega_B$ as the hub at C is rigidly connected to B.

Then, the linear speed at the periphery of pulley B,

$$v_B = \omega_B r_B$$

$$= \omega_C r_B$$

$$= 125 \frac{rad}{sec}(0.35\,ft)$$

$$= 43.75 \frac{ft}{sec}$$

The peripheries of pulley B and D are connected. Therefore, $v_B = v_D$

$$v_B = v_D$$

$$v_B = \omega_D r_D$$

$$\omega_D = \frac{43.75 \frac{ft}{sec}}{r_D}$$

$$= \frac{43.75 \frac{ft}{sec}}{0.2\,ft}$$

$$= 218.75 \frac{rad}{sec}$$

Answers:

Angular velocity of pulley B, $\omega_B = 125 \dfrac{rad}{sec}$

Angular velocity of pulley D, $\omega_D = 218.75 \dfrac{rad}{sec}$

Example 4.7

A motor gives gear A, shown in Figure 4.5, the angular acceleration of $\alpha_A = 1.5 + 0.005\theta^2$ rad/sec^2, where θ is in rads. If this gear is initially turning at 10 rad/sec, determine the angular velocity of gear B after A undergoes an angular displacement of 12 revolutions. Gears A and B have diameters of 200 mm and 100 mm, respectively.

SOLUTION

Given,

Angular acceleration of gear A, $\alpha_A = 1.5 + 0.005\theta^2$ rad/sec^2

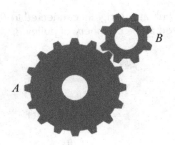

FIGURE 4.5   Gears A and B for Example 4.7

Initial angular velocity of gear A, ω_{Ao} = 10 rad/sec
Angular velocity of gear B, ω_B =?
Angular displacement of gear A, θ_A = 12 rev = 12 rev x 2π rad/rev = 24π
  rad
Radius of gear A, r_A = 200 mm/2 = 100 mm = 0.1 m
Radius of gear B, r_B = 100 mm/2 = 50 mm = 0.05 m
The peripheries of pulleys A and B are connected. Therefore, $v_A = v_B$

$$\omega_A r_A = \omega_B r_B$$

r_A and r_B are known. We need to find out ω_A after 12 revolutions, then ω_B.
Note that, $\alpha = \dfrac{d\omega}{dt}$ or $\int d\omega = \int \alpha dt$ is not appropriate here as no time
  information is given.
The angular acceleration can be expressed as:

$$\alpha = \frac{d\omega}{dt} = \frac{d\omega}{d\theta}\frac{d\theta}{dt} = \frac{d\omega}{d\theta}\omega$$

$$\omega d\omega = \alpha d\theta$$

$$\int_{10\frac{rad}{sec}}^{\omega_A} \omega d\omega = \int_{0}^{\theta=24\pi} \left(1.5 + 0.005\theta^2\right)d\theta$$

$$\left.\frac{\omega^2}{2}\right|_{10}^{\omega_A} = \left[1.5\theta + \frac{0.005}{3}\theta^3\right]_0^{24\pi}$$

$$\frac{\omega_A^2}{2} - \frac{10^2}{2} = \left[1.5(24\pi) + \frac{0.005}{3}(24\pi)^3\right] - [0+0]$$

$$\omega_A = \sqrt{2(50+113+714)}$$

$$\omega_A = 41.9\frac{rad}{sec}$$

$$\omega_A r_A = \omega_B r_B$$

$$\omega_B = \frac{\omega_A r_A}{r_B} = \frac{41.9 \frac{rad}{sec}(0.1m)}{0.05m} = 83.8 \frac{rad}{sec}$$

Answer: Angular velocity of gear B, $\omega_B = 83.8 \frac{rad}{sec}$

Example 4.8

The 6-in diameter disk starts with an angular velocity of 2 rad/sec and is given an angular acceleration of $\alpha = 0.003\theta^{1.5}$ rad/sec^2, where θ is in radians. Determine the magnitude of the linear acceleration after five revolutions.

SOLUTION

Given,

    Radius, $r = 6$ in./2 = 3 in.
    Initial angular velocity, $\omega_o = 2$ rad/sec
    Angular acceleration, $\alpha = 0.003\theta^{1.5}$ rad/sec^2
    Displacement, $\theta = 5$ rev = 5 rev x 2π rad/rev= 10π rad
    Acceleration, $a = ?$

Note that, $\alpha = \frac{d\omega}{dt}$ or $\int d\omega = \int \alpha dt$ is not appropriate here as no time information is given.

The angular acceleration can be expressed as:

$$\alpha = \frac{d\omega}{dt} = \frac{d\omega}{d\theta}\frac{d\theta}{dt} = \frac{d\omega}{d\theta}\omega$$

$$\omega d\omega = \alpha d\theta$$

$$\int_{2\frac{rad}{sec}}^{\omega} \omega d\omega = \int_0^{\theta=10\pi} 0.003\theta^{1.5}d\theta$$

$$\left.\frac{\omega^2}{2}\right|_2^{\omega} = \left.\frac{0.003}{2.5}\theta^{2.5}\right|_0^{10\pi}$$

$$\frac{\omega^2}{2} - \frac{2^2}{2} = \left[\frac{0.003}{2.5}(10\pi)^{2.5}\right] - \left[\frac{0.003}{2.5}(0)^{2.5}\right]$$

$$\omega = 4.2 \frac{rad}{sec}$$

Therefore, angular acceleration,

$$\alpha = 0.003\theta^{1.5} = 0.003(10\pi)^{1.5} = 0.53\frac{rad}{sec^2}$$

Tangential acceleration, $a_t = \alpha r = 0.53\frac{rad}{sec^2}(3.0\,in.) = 1.58\frac{in.}{sec^2}$

Normal acceleration, $a_n = \omega^2 r = \left(4.2\frac{rad}{sec}\right)^2 (3.0\,in.) = 52.92\frac{in.}{sec^2}$

Linear acceleration,

$$a = \sqrt{(a_t)^2 + (a_n)^2} = \sqrt{\left(1.58\frac{in.}{sec^2}\right)^2 + \left(52.92\frac{in.}{sec^2}\right)^2} = 52.94\frac{in.}{sec^2}$$

Answer: 52.94 in./sec²

Note: Only velocity or acceleration means linear velocity or linear acceleration, respectively.

4.4   GENERAL PLANE MOTION

A simultaneous translation and rotation occurs in general plane motion. The
knowledge of angular rotation of a line in the body and the motion of a point on
the body is required to analyze this type of motion. Let us first see how translation
knowledge is used. In Figure 4.6, let X-Y be the original and the fixed coordinate
system and let x-y be the translating coordinate system, and A and B are two
points on the body.

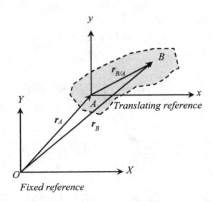

FIGURE 4.6   Reference Axes and the Translating Axes

If $\vec{r}_A$ be the base point in the body and the relative position vector $\vec{r}_{B/A}$ locates point B with respect to point A, then:

$$\vec{r}_B = \vec{r}_A + \vec{r}_{B/A}$$

The velocity can be obtained by differentiating the position vectors as follows:

$$\frac{d}{dt}\left(\vec{r}_B\right) = \frac{d}{dt}\left(\vec{r}_A\right) + \frac{d}{dt}\left(\vec{r}_{B/A}\right)$$

$$\vec{v}_B = \vec{v}_A + \vec{v}_{B/A}$$

$\frac{d}{dt}\left(\vec{r}_{B/A}\right) \neq 0$ as the position of B with respect to A changes due to rotation. As this velocity represents the effect of circular motion about A, it can be written as:

$$\vec{v}_{B/A} = \omega \times \vec{r}_{B/A}$$

Therefore, $\vec{v}_B = \vec{v}_A + \vec{v}_{B/A} = \vec{v}_B = \vec{v}_A + \omega \times \vec{r}_{B/A}$

The acceleration can be determined by differentiating the velocity equation.

$$\frac{d}{dt}\left(\vec{v}_B\right) = \frac{d}{dt}\left(\vec{v}_A\right) + \frac{d}{dt}\left(\vec{v}_{B/A}\right)$$

Now, $\frac{d}{dt}\left(\vec{v}_B\right) = \vec{a}_B$

$$\frac{d}{dt}\left(\vec{v}_A\right) = \vec{a}_A$$

$$\frac{d}{dt}\left(\vec{v}_{B/A}\right) = \vec{a}_{B/A}$$

$$\vec{a}_{B/A} = \left(\vec{a}_{B/A}\right)_t + \left(\vec{a}_{B/A}\right)_n$$

$$\left(\vec{a}_{B/A}\right)_t = \alpha \vec{r}_{B/A}$$

$$\left(\vec{a}_{B/A}\right)_n = -\omega^2 r_{B/A}$$

$\vec{a}_B$ = acceleration of point B.

$\vec{a}_A$ = acceleration of point A.

$\left(\vec{a}_{B/A}\right)_t$ = relative tangential acceleration component of B with respect to A.

$\left(\vec{a}_{B/A}\right)_t = \alpha \vec{r}_{B/A}$ and the direction is perpendicular to $\vec{r}_{B/A}$.

$\left(\vec{a}_{B/A}\right)_n$ = relative normal acceleration component of B with respect to A.

$\left(\vec{a}_{B/A}\right)_n = \omega^2 \vec{r}_{B/A}$ and the direction is from B toward A.

Therefore, $\vec{a}_B = \vec{a}_A + \vec{\alpha} \times \vec{r}_{B/A} - \omega^2 \vec{r}_{B/A}$

Example 4.9

If the block at C moves upward at 3 ft/sec, determine the angular velocity of bar AB at the instant shown in Figure 4.7.

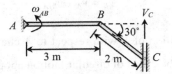

FIGURE 4.7

SOLUTION

The AB link is rotating about a fixed point A; v_B is directed perpendicular to link AB, as shown in Figure 4.8. Therefore, the velocity at B can be written as:

$$\vec{v}_B = \omega_{AB} r_{AB} \vec{k} = \omega_{AB} 3\vec{k} = 3\omega_{AB} \vec{k}$$

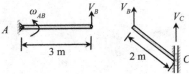

FIGURE 4.8

$$\vec{r}_{C/B} = 2\cos30°\vec{i} - 2\sin30°\vec{j} \text{ m}$$

$$\vec{v}_C = 3\vec{j}\ \frac{\text{m}}{\text{sec}}$$

$$\vec{v}_B = \omega_{AB} r_{AB} \vec{k} = \omega_{AB} 3\vec{k} = 3\omega_{AB} \vec{k}$$

$$\vec{v}_C = \vec{v}_B + \omega_{BC} \times \vec{r}_{C/B}$$

$$3\vec{j} = 3\omega_{AB}\vec{j} + \left(\omega_{BC}\vec{k}\right) \times \left(2\cos30°\vec{i} - 2\sin30°\vec{j}\right)$$

$$3\vec{j} = 3\omega_{AB}\vec{j} + \left(2\cos30°\left(\omega_{BC}\right)\left(\vec{j}\right) - 2\sin30°\left(\omega_{BC}\right)\left(-\vec{i}\right)\right)$$

$$3\vec{j} = 3\omega_{AB}\vec{j} + 1.73\omega_{BC}\vec{j} + \omega_{BC}\,\vec{i}$$

Equating $\vec{i}$ components –

$$\omega_{BC} = 0$$

Equating $\vec{j}$ components –

$$3\vec{j} = 3\omega_{AB}\vec{j} + 1.73(0)\vec{j}$$

$$\omega_{AB} = 1.0\,\frac{\text{rad}}{\text{sec}}$$

Answer: $\omega_{AB} = 1.0\,\dfrac{\text{rad}}{\text{sec}}$

Example 4.10

The angular velocity of bar AB is given as 5 rad/sec. Determine the velocity of the slider-block C at the instant shown in Figure 4.9.

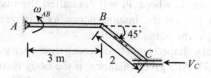

FIGURE 4.9

SOLUTION

The AB link is rotating about fixed point A. As a result, v_B is always directed perpendicular to link AB, and its magnitude is $v_B = \omega_{AB}r_{AB} = \left(5\,\dfrac{\text{rad}}{\text{sec}}\right)(3\text{m}) = 15\,\dfrac{\text{m}}{\text{sec}}$.

At the instant shown in Figure 4.10, v_B is directed perpendicular to the horizontal. Also, block C is moving horizontally due to the constraint of the guide.

$$r_{B/IC} = r_{C/IC} = (2\text{m})\cos45 = 1.414\text{ m}$$

The angular velocity of bar BC can be obtained as $\omega_{BC} = \dfrac{v_B}{r_{B/IC}} = \dfrac{15\,\dfrac{\text{m}}{\text{sec}}}{1.414} = 10.6\,\dfrac{\text{rad}}{\text{sec}}$

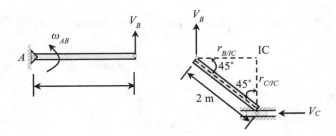

FIGURE 4.10

Thus, the velocity of block C is $v_C = \omega_{BC} r_{C/IC} = \left(10.6\ \dfrac{\text{rad}}{\text{sec}}\right)(1.414\ \text{m}) = 15\ \dfrac{\text{m}}{\text{sec}}$

Answer: $v_C = 15\ \dfrac{\text{m}}{\text{sec}}$

4.5 ROTATION ABOUT INSTANTANEOUS AXIS

In the previous section, it was shown that for general planar motion, the velocity of point B could be found as:

$$\vec{v}_B = \vec{v}_A + \vec{v}_{B/A}$$

If point A can be found such that $\vec{v}_A = 0$, then, $\vec{v}_B = \vec{v}_{B/A} = \vec{\omega} \times \vec{r}_{B/A}$. For a general plane motion, point A, where $\vec{v}_A = 0$, is called the instantaneous center of zero velocity (IC). It lies on the instantaneous axis of zero velocity, which is always perpendicular to the plane of motion.

Consider an arbitrary shaped rigid body going through a general planar motion, as shown in Figure 4.11. At a given instance, if the body is hinged at point IC and is rotating about this point, then point IC is known as the instantaneous center of zero velocity. Thus, the velocity at point IC at that instance will be zero. If A is

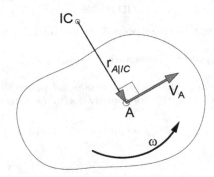

FIGURE 4.11 Instantaneous Centers of Zero Velocity

an arbitrary point on this plane of motion, then the position vector $\overline{r}_{AIC}$ will be normal to the velocity vector $\overline{V}_A$. The magnitude of the angular velocity ω can be determined using $\dfrac{V}{r_{AIC}}$.

The location of the instantaneous center of zero velocity for a rigid body undergoing general planar motion can be found based on the magnitudes of velocities at different locations and their corresponding directions. Consider two particles A and B on a planar motion, as shown in Figure 4.12 (a), with the corresponding velocities of V_A and V_B, where the directions of V_A and V_B are non-parallel. If two lines are drawn at point A and point B such that they are normal to the corresponding velocity directions, they will intersect at a point. This intersection point IC is the center of instantaneous zero velocity.

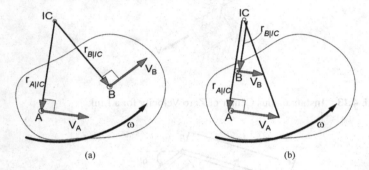

(a)   (b)

FIGURE 4.12   Finding the Instantaneous Centers of Zero Velocities

However, if the directions of the velocities at point A and point B are parallel to each other, as shown in Figure 4.12 (b), the instantaneous center can be found by intersecting line AB with the line that joins the tips of vector V_A and vector V_B. The lengths of V_A and V_B must be scaled based on the corresponding magnitudes of the velocities.

To understand how the concept of the instantaneous center of zero velocity works, consider the following mechanism as shown in Figure 4.13, where the link AB is undergoing a general planar motion. First, identify the directions of motion for slider blocks A and B. Now draw lines at point A and point B normal to the corresponding velocity directions. These two lines intersect at point IC. Thus IC is the instantaneous center of zero velocity at the given instant. If the magnitude of V_A is known, the angular velocity ω can be obtained as $\omega = \dfrac{V_A}{r_{AIC}}$. Also, the velocity V_B can be obtained as $V_B = \omega . r_{BIC}$.

Example 4.11

The angular velocity of bar *AB* is given as 5 rad/sec. Determine the velocity of the slider-block *C* at the instant shown in Figure 4.14.

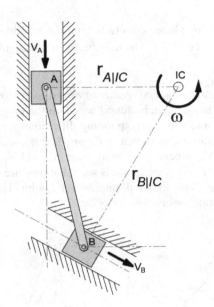

FIGURE 4.13 Instantaneous Center of Zero Velocity for a Link

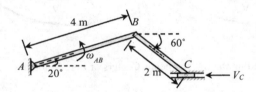

FIGURE 4.14

SOLUTION

The link AB is rotating about the fixed point A, so v_B is always directed perpendicular to link AB, and its magnitude is $v_B = \omega_{AB} r_{AB} = \left(5\,\frac{\text{rad}}{\text{sec}} \right)(4\,\text{m}) = 20\,\frac{\text{m}}{\text{sec}}$.

At the instant shown in Figure 4.15, v_B is directed perpendicular to the horizontal. Also, block C is moving horizontally due to the constraint of the guide.

$$\frac{r_{C/IC}}{\sin 80°} = \frac{2\,\text{m}}{\sin 70°}$$

$$r_{C/IC} = \left(\frac{2\,\text{m}}{\sin 70°} \right) \sin 80° = 2.1\,\text{m}$$

$$\frac{r_{B/IC}}{\sin 30°} = \frac{2\,\text{m}}{\sin 70°}$$

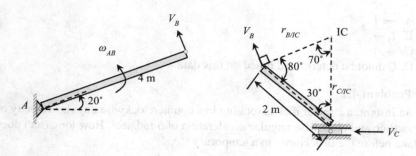

FIGURE 4.15

$$r_{B/IC} = \left(\frac{2m}{\sin70°}\right)\sin30° = 1.06 \text{ m}$$

The angular velocity of bar *BC* is given by $\omega_{BC} = \dfrac{v_B}{r_{B/IC}} = \dfrac{20\,\dfrac{m}{\sec}}{1.06m} = 18.87\,\dfrac{rad}{\sec}$

Thus, the velocity of block *C* is

$$v_C = \omega_{BC}r_{C/IC} = \left(18.87\,\frac{rad}{\sec}\right)(2.1m) = 39.6\,\frac{m}{\sec}$$

Answer: $v_C = 39.6\,\dfrac{m}{\sec}$

FUNDAMENTALS OF ENGINEERING (FE) EXAM STYLE QUESTIONS

FE Problem 4.1

A particle rotates in a circular path at a constant speed in a counterclockwise direction. Consider a time interval during which the particle moves along this circular path from point *P* to point *Q*. Point *Q* is exactly halfway around the circle from point *P*, as shown in Figure 4.16. The direction of the average linear velocity and the average tangential acceleration, respectively, during this time interval, are most nearly:

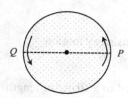

FIGURE 4.16

A. ←, ↓
B. ↑, →
C. ←, ↑
D. Cannot be determined based on this data

FE Problem 4.2

At an instant, a circular disk is rotating in a counterclockwise angular velocity of 10 rad/sec and clockwise angular acceleration of 5 rad/sec². How long (sec) does it take before the disk comes to a temporary stop?

A. 2
B. 4
C. 6
D. 8

FE Problem 4.3

Three forces act on an object. Identify the correct statement below if the object is in translational equilibrium.
[select two options]

A. the vector sum of all three forces must equal zero
B. two of the forces must be perpendicular
C. the magnitudes of each of the three forces must be equal
D. each force has a magnitude of zero
E. all three forces must be parallel

FE Problem 4.4

A motor flywheel goes from rest to 1,000 rpm in 6 sec. What is the approximate number of revolutions made by the flywheel?

A. 25
B. 50
C. 100
D. 250

FE Problem 4.5

When a body moves around a fixed axis, it has

A. a rotary motion
B. a circular motion
C. a translatory motion
D. a rotary motion and translatory motion

FE Problem 4.6

A stone of mass of 1 kg is tied to a string of length 1 m and spun in a circle on the horizontal plane at a constant angular speed of 5 rad/sec. The tension (N) in the string is:

A. 5
B. 10
C. 15
D. 25
E. none of these options

FE Problem 4.7
The motion of a bike wheel is most nearly:

A. translatory
B. rotary
C. rotary and translatory
D. curvilinear

FE Problem 4.8
The point about which combined motion of rotation and translation of a rigid body takes place is known as:

A. virtual center
B. instantaneous center
C. instantaneous axis
D. point of rotation
E. all of these options

FE Problem 4.9
The velocity of a point on the rim of a wheel is given as 50 m/sec. If the diameter of the rim is given as 10 m, what will be the angular velocity (rad/sec) of the wheel?

A. 20
B. 15
C. 10
D. 5

FE Problem 4.10
For the translation of a rigid body, which of the following is not correct?

A. $\vec{a}_A = \vec{a}_B$

B. $\vec{v}_A = \vec{v}_B$

C. $\alpha\, d\theta = \omega\, d\omega$

D. $\dfrac{d}{dt}\left(\vec{r}_{A/B}\right) = 0$

PRACTICE PROBLEMS

Problem 4.1

An electric motor is rotating at 322 rad/sec² starting from rest. Calculate the angular displacement of the motor after 2 min.

Problem 4.2

An electric motor is rotating at 322 rad/sec² starting from rest. Calculate the angular velocity after 1,000 revolutions.

Problem 4.3

Bar *AO*, with a length $r = 100$ mm, is rotating at a constant angular acceleration of 0.5 rad/s² in a counterclockwise direction, as shown in the image below. What is the velocity of the bar at point A after 8 sec if the initial angular displacement and the initial velocity were both zero? Also, what is the angular displacement after 8 sec? (Figure 4.17)

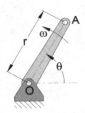

FIGURE 4.17

Problem 4.4

A wheel starts with an initial clockwise angular velocity of 20 rad/sec and with a constant angular acceleration of 2 rad/sec². How many revolutions will it undergo to acquire a clockwise angular velocity of 80 rad/sec? Also, calculate the time required to complete these revolutions.

Problem 4.5

The angular position of a car traveling around a curve is described by $\theta(t) = 15t + 2t^2 - 6t^3 + 5$ radians, where *t* is in sec. Calculate the linear acceleration of the car at the beginning and after 25 sec.

Problem 4.6

Wheel-A started spinning about its center O in a counterclockwise direction with an angular acceleration of $\alpha = 2\theta + \pi$. Express the angular velocity ω in terms of angular displacement θ. What would be the angular velocity ω if the angular displacement θ is given as 2.7 radians? (Figure 4.18)

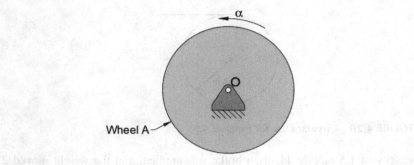

FIGURE 4.18

Wheel A

Problem 4.7

The angular velocity of a 4-in diameter disk, shown in Figure 4.19, is defined by $\omega = 2t^3 + 3$ rad/sec, where t is in sec. Determine the magnitudes of velocity and acceleration at A and B of the disk when $t = 0.5$ sec.

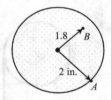

FIGURE 4.19   Circular Disk for Problems 4.7 and 4.8

Problem 4.8

The disk, shown in Figure 4.19, is initially rotating at 12 rad/sec. If it is subjected to a constant angular acceleration of $\alpha = 4$ rad/sec^2, determine the magnitudes of the velocity and acceleration of points A and B when the disk undergoes 12 revolutions.

Problem 4.9

The cord, wrapped around the disk as shown in Figure 4.20, is given an acceleration of $a = (6t^2)$ ft/sec^2, where t is in sec. Starting from rest, determine the angular displacement, angular velocity, and angular acceleration of the disk when $t = 10$ sec.

Problem 4.10

A weight is attached to one end of a cord, while the other end is wrapped around the cylinder, as shown below. The weight was moving downward at a uniform

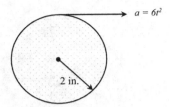

FIGURE 4.20   Circular Disk for Problem 4.9

velocity of 1.5 m/s. Suddenly a brake was applied, and the weight moved 2 m downward before coming to a stop. Find the angular acceleration of the cylinder and the time required to stop its rotation. The diameter of the cylinder is given as 0.75 m (Figure 4.21).

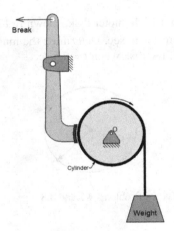

FIGURE 4.21

Problem 4.11

The angular acceleration of a 0.14-m diameter circular disk can be represented as $\alpha = 15t^2 - 4t + 2$ rad/sec², where t is in sec. The initial angular velocity of the disk was 6 rad/sec. Determine the following after 5 sec of rotation:

(a) the linear velocity
(b) the linear acceleration

Problem 4.12

The power of a car engine is transmitted using the belt-and-pulley arrangement shown in Figure 4.22. If the engine turns pulley B at 50 rad/sec, determine the angular velocities of pulley A and pulley D. The hub at C is rigidly connected to B and turns with it. The diameters of pulley A, B, C, and D are 0.5 ft, 0.7 ft, 0.2 ft, and 0.4 ft, respectively.

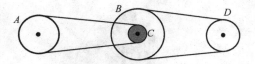

FIGURE 4.22   Belt-and-Pulley Arrangement for Problem 4.12

Problem 4.13

In the following belt-pulley system, as shown in the image below, pulley A is turning at a rate of 35 rad/s. Given that pulley B and pulley C are joined together, and pulleys D and E are joined together. The diameters for pulleys A, B, C, D, E, and F are given as 3 in., 5 in., 2 in., 4 in., 2 in., and 3.5 in., respectively. Determine the rpm for pulley F (Figure 4.23).

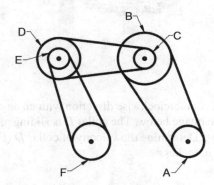

FIGURE 4.23

Problem 4.14

A 4-in. diameter disk starts with an angular velocity of 1 rad/sec and is given an angular acceleration of $\alpha = 0.025\theta^{1.25}$ rad/sec², where θ is in radians. Determine the magnitudes of the normal and tangential components of acceleration of a point on the rim (periphery) after five revolutions.

Problem 4.15

In the following power transmission system, as shown in the image below, gear A is turning at a rate of 25 rad/s. The pitch diameters for gear A, B, C, and D are given as 200 mm, 80 mm, 150 mm, and 60 mm, respectively. What is the angular velocity of gear D? (Figure 4.24)

Problem 4.16

A 4-in. diameter disk starts with an angular velocity of 1 rad/sec and is given an angular acceleration of $\alpha = 2.5t^{1.25}$ rad/sec², where θ is in radians. Determine the magnitudes of the normal and tangential components of acceleration of a point on the rim (periphery) after 5 sec.

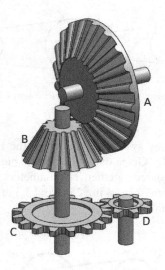

FIGURE 4.24

Problem 4.17

Disk A is rotating in a counterclockwise direction with an angular velocity of $\omega =$ 6 rad/s as shown in the image below. The collar D is sliding along the vertical rod while disk A is rotating. Determine the velocity of collar D (Figure 4.25). Length dimensions are given in inches.

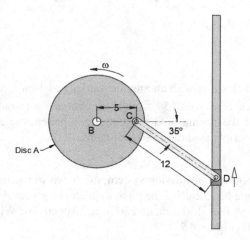

FIGURE 4.25

Problem 4.18

The sun gear S in the planetary gear system, as shown below, is rotating in a counterclockwise direction at an angular speed of 10 rad/s. If the ring gear R is fixed, determine the angular velocities of the planet gears P and the planet-carrier C. Given that the diameters for ring gear, planet gear, and sun gear are 90 mm, 25 mm, and 40 mm, respectively (Figure 4.26).

FIGURE 4.26

Problem 4.19

Two bars, AB and BC, are connected, as shown in Figure 4.27. If the block at C is moving in a downward direction at 4 ft/sec, find out the angular velocity of bar AB at the instant shown.

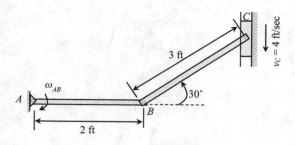

FIGURE 4.27   The Bar Arrangement for Problem 4.19

Problem 4.20

Disk A has an angular velocity of 8 rad/s in a counterclockwise direction, at the instant shown in the image below. What is the velocity of block D? Given that L = 18 in. and r = 4.5 in. (Figure 4.28).

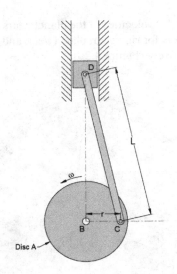

FIGURE 4.28

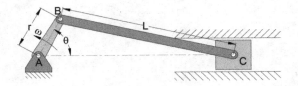

FIGURE 4.29

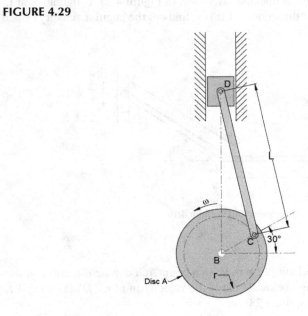

FIGURE 4.30

Problem 4.21

Bar AB with a length of 5 in. is rotating counterclockwise about point A with an angular velocity of 7.5 rad/s, as shown in the image below. The length of bar CD is $L = 20$ in. Determine the velocity of block C at the instant when $\theta = 60$ (Figure 4.29).

Problem 4.22

At the given instant, block D is moving upward with a velocity of 3.5 m/s. L and r are given as 750 mm and 200 mm, respectively. Determine the angular velocity of disk A (Figure 4.30).

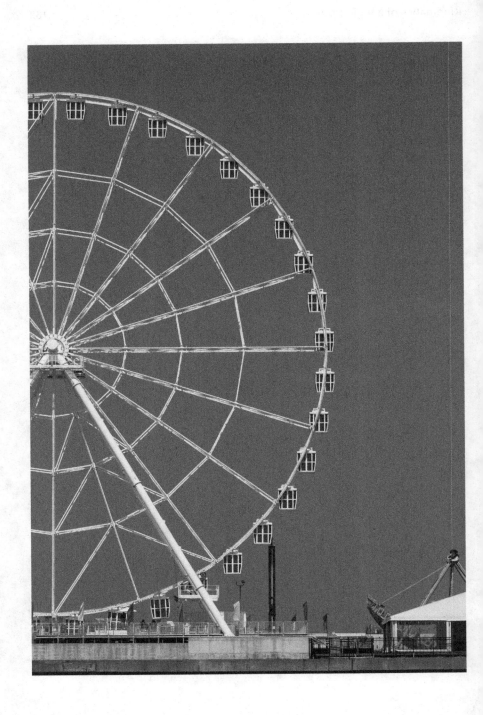

5 Kinetics of a Rigid Body

5.1 GENERAL

Rigid body kinetics studies the movement of systems of interconnected bodies under the action of external forces. The assumption that the bodies are rigid, which means that they do not deform under the action of applied forces, simplifies the analysis by reducing the parameters that describe the configuration of the system to the translation and rotation of reference frames attached to each body.

The dynamics of a rigid body system is described by the laws of kinematics and by the application of Newton's law (kinetics). The solution of these equations of motion provides a description of the position, velocity, acceleration of the individual components of the system, energy, work, etc. The formulation and solution of rigid body kinetics is a crucial tool in the simulation of mechanical systems. This chapter discusses how Newton's Law of Motion can be applied for rigid body motion.

5.2 MASS MOMENT OF INERTIA

Mass moment of inertia is important in rigid body kinetics as it measures the resistance of a body to angular acceleration. The larger the mass moment of inertia, the larger the resistance of the body to an angular acceleration. When a rigid body is confined in a fixed axis, then (if force is applied on that body) it cannot move in a straight line due to confinement. The body rotates around the axis and there is an angular displacement of each of the particles of the body. This type of motion with respect to an axis is called rotational motion. The axis can be either inside or outside the body.

If a rigid body rotates around an axis, the moment of inertia of that body with respect to that axis means the summation of the product of square of distance from the axis and the mass of each of the particles of the body. Mathematically, the mass moment of inertia can be expressed in the following way:

$$I = \int r^2 dm$$

where dm is the mass of an infinitesimally small element of the body at distance r from the axis. Mass moment of inertia does not depend on angular velocity of particles but depends instead on the distribution of particles about the axis of rotation.

In the Cartesian coordinate system, the mass moment of inertia can be defined as:

DOI: 10.1201/9781003283959-5

Mass moment of inertia about the x-axis: $I_x = \sum y^2 \Delta m$

Mass moment of inertia about the y-axis: $I_y = \sum x^2 \Delta m$

The unit of mass moment of inertia is kg.m² or slug.ft². The parallel axis theorem can be applied to the above equations to determine the mass moment of inertia at any parallel axis if the mass moment of inertia at the centroidal axis (I_{xc}) is known.

$$I_x = \sum I_{xc} + \sum md^2$$

where:

m = mass of the body

d = parallel distance from the mass centroid to any parallel axis

$d = \bar{y} - y_i$ or $d = y_i - \bar{y}$

$\bar{y}$ = mass centroid along the y-axis

y_i = mass centroid of individual component of the body

Let us see how calculus is used do define the above-mentioned terms. Figure 5.1 shows an arbitrary body with an infinitesimal mass of dm.

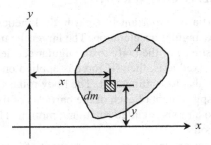

FIGURE 5.1   An Arbitrary Body in a Cartesian Coordinate System

The mass moment of inertia of dA about x-axis can be written as, $I_x = \int y^2 dm$

The mass moment of inertia of dA about y-axis can be written as, $I_y = \int x^2 dm$

The polar mass moment of inertia can be written as, $J = \int r^2 dm = \int (x^2 + y^2) dm$

For example, the mass moment of inertia of a thin uniform rod is determined here about an axis through its center and perpendicular to its length. Let AB be a thin uniform rod of length L and mass m, free to rotate about the axis CD which passes through the center and perpendicular to the length of the rod, as shown in Figure 5.2. Since the rod is uniform, the mass per unit length = m/L. Therefore, at

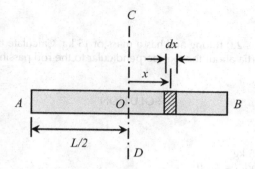

FIGURE 5.2   A Thin Uniform Rod

a distance, x, from the axis, CD, let the dx be a very small length whose mass is $dm = \dfrac{m}{L} dx$. As the dx is very small, all the particles in the dx can be considered to be at same distance from CD. Now, the mass moment of inertia about the CD axis,

$$I_{CD} = \int_{-L/2}^{L/2} x^2 dm = \int_{-L/2}^{L/2} x^2 \frac{m}{L} dx = \frac{m}{L} \int_{-L/2}^{L/2} x^2 dx = \frac{m}{L} \left[\frac{x^3}{3} \right]_{-L/2}^{L/2} = \frac{mL^2}{12}$$

5.3   RADIUS OF GYRATION

If the total mass of a body is considered to be concentrated at a fixed point, so that the mass moment of inertia of that point with respect to a fixed axis is equal to the total moment of inertia of that rigid body about that axis, then the distance of that point from the axis is called radius of gyration (r). It can be expressed as:

$$r = \sqrt{\frac{I}{m}}$$

The unit of radius of gyration (r) is m or ft. The radius of gyration (r) in the Cartesian coordinate system can be presented as:

The radius of gyration about the x-axis, $r_x = \sqrt{\dfrac{I_x}{m}}$

The radius of gyration about the y-axis, $r_y = \sqrt{\dfrac{I_y}{m}}$

The polar radius of gyration, $r_j = \sqrt{\dfrac{J}{m}}$

Example 5.1

A uniform rod is 2.0 ft long and has a mass of 15 kg. Calculate the rod's mass moment of inertia about the axis perpendicular to the rod passing through the centroid.

SOLUTION

Given,

Mass, $m = 15$ kg
Length of rod, $L = 2.0$ ft

From Table 5.1, mass moment of inertia,

$$I_{yc} = I_{zc} = \frac{ml^2}{12} = \frac{(15\,\text{kg})(2.0\,\text{ft})^2}{12} = 5.0 \text{ kg.ft}^2$$

Note: Be careful about which equation to pick. The moment of inertia depends on the reference axis. For example, if the question asks about the axis perpendicular to the rod passing through the end, then we would select,

$$I_y = I_z = \frac{ml^2}{3}.$$

Answer: 5.0 kg.ft²

Example 5.2

A uniform rod is 2.0 ft long and has a mass of 15 kg. Calculate the rod's mass moment of inertia about the axis perpendicular to the rod passing through the edge.

SOLUTION

Given,

Mass, $m = 15$ kg
Length of rod, $L = 2.0$ ft

From Table 5.1, mass moment of inertia,

$$I_{yc} = I_{zc} = \frac{ml^2}{3} = \frac{(15\,\text{kg})(2.0\,\text{ft})^2}{3} = 20.0 \text{ kg.ft}^2$$

Answer: 20.0 kg.ft²

Example 5.3

A hollow cylinder has an inner radius of 2.0 ft, an outer radius of 2.5 ft, and a weight of 150 lb. Calculate the rod's mass moment of inertia about the longitudinal axis passing through the centroid.

TABLE 5.1
Mass Moment of Inertia of Some Common Sections

Figure	Mass & Centroid	Mass Moment of Inertia	(Radius of Gyration)^2
	$M = \rho L A$ $x_c = L/2$ $Y_c = 0$ $Z_c = 0$ $A =$ cross-sectional area of rod $\rho =$ mass/vol.	$I_x = I_{x_c} = 0$ $I_{y_c} = I_{z_c} = ML^2/12$ $I_y = I_z = ML^2/3$	$r_x^2 = r_{x_c}^2 = 0$ $r_{y_c}^2 = r_{z_c}^2 = L^2/12$ $r_y^2 = r_z^2 = L^2/3$
	$M = \rho_c A$ $x_c = R =$ mean radius $Y_c = R =$ mean radius $Z_c = 0$ $A =$ cross-sectional area of ring $\rho =$ mass/vol.	$I_{x_c} = I_{y_c} = MR^2/2$ $I_{z_c} = MR^2$ $I_x = I_y = 3MR^2/2$ $I_z = 3MR^2$	$r_{x_c}^2 = r_{y_c}^2 = R^2/2$ $r_{z_c}^2 = R^2$ $r_x^2 = r_y^2 = 3R^2/2$ $r_z^2 = 3R^2$
	$M = \pi R^2 \rho h$ $x_c = 0$ $Y_c = h/2$ $Z_c = 0$ $\rho =$ mass/vol.	$I_{x_c} = I_{z_c} =$ $M(3R^2 + h^2)/12$ $I_{y_c} = I_y = MR^2/2$ $I_x = I_z = M(3R^2 + 4h^2)/12$	$r_{x_c}^2 = r_{y_c}^2 =$ $(3R^2/h^2)/12$ $r_{y_c}^2 = r_y^2 = R^2/2$ $r_x^2 = r_z^2 =$ $(3R^2 + 4h^2)/12$
	$M = \pi(R_1^2 - R_2^2) \rho h$ $x_c = 0$ $Y_c = h/2$ $Z_c = 0$ $\rho =$ mass/vol.	$I_{x_c} = I_{z_c}$ $= M(3R_1^2 + 3R_2^2 + h2)/12$ $I_{y_c} = I_y = M(R_1^2 + R_2^2)/2$ $I_x = I_z$ $= M(3R_1^2 + 3R_2^2 + 4h^2)/12$	$r_{x_c}^2 = r_{z_c}^2 = (3R_1^2 + 3R_2^2 + h^2)/12$ $r_{y_c}^2 = r_y^2 = (R_1^2 + R_2^2)/2$ $r_x^2 = r_z^2$ $= (3R_1^2 + 3R_2^2 + h^2)/12$
	$M = \dfrac{4}{3} \pi R^3 \rho$ $x_c = 0$ $Y_c = 0$ $Z_c = 0$ $\rho =$ mass/vol.	$I_{x_c} = I_x = 2MR^2/5$ $I_{y_c} = I_y = 2MR^2/5$ $I_{z_c} = I_z = 2MR^2/5$	$r_{x_c}^2 = r_x^2 = 2R^2/5$ $r_{y_c}^2 = r_y^2 = 2R^2/5$ $r_{z_c}^2 = r_z^2 = 2R^2/5$

(Continued)

TABLE 5.1 (CONTINUED)
Mass Moment of Inertia of Some Common Sections

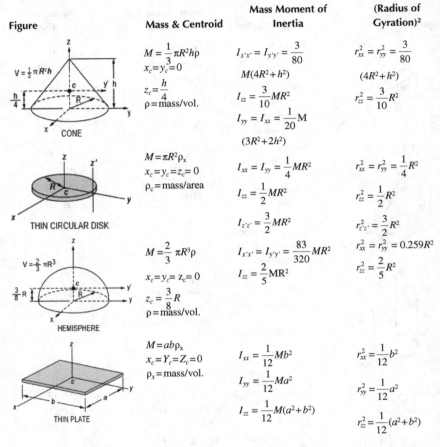

Figure	Mass & Centroid	Mass Moment of Inertia	(Radius of Gyration)^2
CONE $V=\frac{1}{3}\pi R^2 h$	$M=\frac{1}{3}\pi R^2 h\rho$ $x_c=y_c=0$ $z_c=\frac{h}{4}$ $\rho=$ mass/vol.	$I_{x'x'}=I_{y'y'}=\frac{3}{80}$ $M(4R^2+h^2)$ $I_{zz}=\frac{3}{10}MR^2$ $I_{yy}=I_{xx}=\frac{1}{20}M$ $(3R^2+2h^2)$	$r_{xx}^2=r_{yy}^2=\frac{3}{80}$ $(4R^2+h^2)$ $r_{zz}^2=\frac{3}{10}R^2$
THIN CIRCULAR DISK	$M=\pi R^2\rho_x$ $x_c=y_c=z_c=0$ $\rho_c=$ mass/area	$I_{xx}=I_{yy}=\frac{1}{4}MR^2$ $I_{zz}=\frac{1}{2}MR^2$ $I_{z'z'}=\frac{3}{2}MR^2$	$r_{xx}^2=r_{yy}^2=\frac{1}{4}R^2$ $r_{zz}^2=\frac{1}{2}R^2$ $r_{z'z'}^2=\frac{3}{2}R^2$
HEMISPHERE $V=\frac{2}{3}\pi R^3$	$M=\frac{2}{3}\pi R^3\rho$ $x_c=y_c=z_c=0$ $z_c=\frac{3}{8}R$ $\rho=$ mass/vol.	$I_{x'x'}=I_{y'y'}=\frac{83}{320}MR^2$ $I_{zz}=\frac{2}{5}MR^2$	$r_{xx}^2=r_{yy}^2=0.259R^2$ $r_{zz}^2=\frac{2}{5}R^2$
THIN PLATE	$M=ab\rho_x$ $x_c=Y_c=Z_c=0$ $\rho_x=$ mass/vol.	$I_{xx}=\frac{1}{12}Mb^2$ $I_{yy}=\frac{1}{12}Ma^2$ $I_{zz}=\frac{1}{12}M(a^2+b^2)$	$r_{xx}^2=\frac{1}{12}b^2$ $r_{yy}^2=\frac{1}{12}a^2$ $r_{zz}^2=\frac{1}{12}(a^2+b^2)$

Ref: Housner, George W., and Donald E. Hudson, Applied Mechanics Dynamics, D. Van Nostrand Company, Inc., Princeton, NJ, 1959.

SOLUTION

Given,

Weight of cylinder, $W=150$ lb

Mass of cylinder, $m=\dfrac{W}{g}=\dfrac{150 \text{ lb}}{32.2\dfrac{\text{ft}}{\text{sec}^2}}=4.66$ slug

Inner radius, $r_2=2.0$ ft

Outer radius, $r_1 = 2.5$ ft

From Table 5.1, mass moment of inertia, $I_{yc} = I_y = \dfrac{m\left(r_1^2 + r_2^2\right)}{2}$

$$= \dfrac{4.66\,\text{slug}\left((2.5\,\text{ft})^2 + (2.0\,\text{ft})^2\right)}{2}$$

$$= 23.9 \text{ slug.ft}^2$$

Answer: 23.9 slug.ft²
The equation of the mass moment of inertia uses mass, not weight. Therefore, the unit, lb, must be converted to slug for the US customary unit.

Example 5.4

A thin metal plate, shown in Figure 5.3, has the following dimensions with a 1.5-ft diameter hole at its center. The metal weighs 100 pound per square feet (psf).

(a) Calculate the moment of inertia of mass of the plate about an axis perpendicular to the page and passing through the centroid.
(b) Calculate the moment of inertia of mass of the plate about an axis perpendicular to the page and passing through the hanging corner.

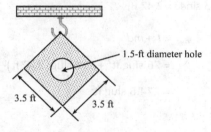

FIGURE 5.3 A Thin Metal Plate for Example 5.4

SOLUTION

Metal density = 100 psf
Metal mass = 100 psf/32.2 ft/sec² = 3.1 ssf (slug per square feet)
Hole radius, $r = 1.5$ ft/2 = 0.75 ft
Plate size ignoring the hole = 3.5 ft × 3.5 ft

Being symmetric, the centroid is located at its center.

Mass of whole plate ignoring hole = (3.5 ft × 3.5 ft) × 3.1 ssf = 37.98 slug
Mass of metal equivalent to hole area = $\pi r^2 \times 3.1$ ssf = π (0.75 ft)² × 3.1 ssf = 5.48 slug

Let the whole plate ignoring the hole be '1' and the hole itself be '2'.

(a) Moment of inertia of mass about the centroid

$$\bar{I} = I_{plate} - I_{hole}$$

From Table 5.1, $\bar{I} = \dfrac{1}{12}m_1\left(a^2 + b^2\right) - \dfrac{1}{2}m_2r_2^2$

$$= \frac{1}{12}(37.98\text{ slug})\left(\left(3.5\,\text{ft}\right)^2 + \left(3.5\,\text{ft}\right)^2\right) - \frac{1}{2}(5.48\text{ slug})\left(0.75\,\text{ft}\right)^2$$

$$= 76\text{ slug.ft}^2$$

(b) Moment of inertia of mass about the hanging corner

To transfer the mass moment of inertia from the center to corner, we need to use the parallel axis theorem.

$$I_{corner} = \bar{I} + md^2$$

Mass, $m = 37.98$ slug $- 5.48$ slug $= 32.5$ slug
Distance to transfer, $d =$ center to corner

$$= \text{half of the diagonal} = \frac{1}{2}\sqrt{\left(3.5\,\text{ft}\right)^2 + \left(3.5\,\text{ft}\right)^2} = 2.47\text{ ft}$$

[Or, $d = 3.5\ \sin 45 = 2.47$ ft]

$$I_{corner} = \bar{I} + md^2$$

$$= 76\text{ slug.ft}^2 + 32.5\,\text{slug}\left(2.47\,\text{ft}\right)^2$$

$$= 274.6\text{ slug.ft}^2$$

Answers:
(a)   *76 slug.ft²*
(b)   *274.6 slug.ft²*

Again, to remind you, in the US customary unit, lbf (simply lb) is the unit of weight; lbm (slug) is the unit of mass.

Example 5.5

The pendulum, shown in Figure 5.4, consists of a 3-kg slender rod and a 5-kg thin plate. Determine the location of the center of mass, G, of the pendulum; then calculate the moment of inertia of mass of the pendulum about an axis perpendicular to the page and passing through G.

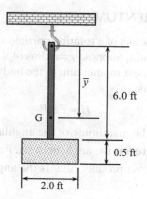

FIGURE 5.4 The Pendulum for Example 5.5

SOLUTION

Let the rod be '1' and the plate be '2'.

Mass of rod, $m1 = 3$ kg
Mass of thin plate, $m2 = 5$ kg

Let us use the hook as the reference point.

Centroid of rod from the hook, $y_1 = 6.0$ ft/2 $= 3.0$ ft
Centroid of thin plate from the hook, $y_2 = 6.0$ ft $+ 0.5$ ft/2 $= 6.25$ ft

Combined centroid from the hook,

$$\bar{y} = \frac{y_1 m_1 + y_2 m_2}{m_1 + m_2} = \frac{(3.0\,\text{ft})(3\text{kg}) + (6.25\,\text{ft})(5\text{kg})}{3 + 5\ \text{kg}} = 5.03\,\text{ft}$$

Mass moment of inertia, $I_G = \sum \left(\bar{I_G} + md^2 \right)$

$$I_G = I_{rod} + I_{plate}$$

We will determine the mass moment of inertia of the rod about its own centroid and then transfer it to G. Similarly, we will determine the mass moment of inertia of the plate about its own centroid and then transfer it to G.

$$I_G = \left[\frac{1}{12}m_1 L_1^2 + m_1 (\bar{y} - y_1)^2 \right] + \left[\frac{1}{12}m_2 (a^2 + b^2) + m_2 (\bar{y} - y_2)^2 \right]$$

$$= \frac{1}{12}(3\text{kg})(6\,\text{ft})^2 + (3\text{kg})(5.03\,\text{ft} - 3\,\text{ft})^2 + \frac{1}{12}(5\text{kg})\left((2\,\text{ft})^2 + (0.5\,\text{ft})^2\right)$$

$$+ (5\text{kg})(5.03\,\text{ft} - 6.25\ \text{ft})^2 = 30.58\,\text{kg.ft}^2$$

Answers: 5.03 ft from the hook, 30.58 kg.ft²

5.4 ANGULAR MOMENTUM

The vector product of the radius of a rotating particle and the linear momentum is called the angular momentum. Suppose r=radius of a particle with respect to the center of rotation and P=linear momentum of the body, then the angular momentum (H) can be expressed as:

$$\vec{H} = \vec{r} \times \vec{P}$$

It is a vector quantity and the magnitude of the angular momentum can be determined as $H = rP \sin\theta$, where θ is the angle between $\vec{r}$ and $\vec{P}$. The direction of H is to be determined by the cross-product rule. If the angle between $\vec{r}$ and $\vec{P}$ is 90°, then:

$$H = rP \sin\theta = rP = r(mv) = rm(\omega r) = \omega(mr^2) = \omega I = I\omega$$

where, ω=angular velocity and mass moment of inertia, $I = mr^2$

Therefore, it can be said that the angular momentum of an object is the product of its moment of inertia about the axis of rotation and its angular velocity.

Example 5.6

The mass of a metal sphere is 6 g. It is rotated 4 times per sec by fastening it at the end of a thread of 3 m length. Calculate its angular momentum about the axis passing through the center and perpendicular to the circular path.

SOLUTION

Given,

Mass, m=6 g=0.006 kg
Radius of rotation, r=3 m
Frequency, f=4 Hz

As the axis is perpendicular to the circular path, angular momentum, $H = I\omega = mr^2\omega = mr^2(2\pi f)$

$$= (0.006\,\text{kg})(3\,\text{m})^2 (2\text{\AA})(4\,\text{Hz})$$

$$= 1.356 \text{ kg.m}^2\text{sec}^{-1}$$

Answer: 1.356 kg.m²/sec

Example 5.7

A boy weighing 40 kg is revolving at an angular velocity of 6 rpm on a merry-go-round around a circular path of 20 m diameter. Determine the angular momentum of the boy.

SOLUTION

Angular velocity, $\omega = 6\,\text{rpm} = \dfrac{6(2\pi)}{60\,\text{sec}} = \dfrac{\pi\;\text{rad}}{5\;\text{sec}}$

Mass, $m = 40$ kg

Radius of rotation, $r = 20$ m/2 $= 10$ m

Angular momentum, $H = mr^2\omega = 40\,\text{kg}\,(10\,\text{m})^2\left(\dfrac{\pi}{5}\,\dfrac{\text{rad}}{\text{sec}}\right) = 2{,}512\ \text{kg.m}^2\,\text{sec}^{-1}$

Answer: 2,512 kg.m²/sec

Example 5.8

A body of mass 500 g is revolving in a circular path of radius 2 m. If the time period of revolution is 10 sec, calculate the angular momentum of the body.

SOLUTION

Mass, $m = 500$ g $= 0.5$ kg

Radius of rotation, $r = 2$ m

Time period, $\tau = 10$ sec

Angular momentum, $H = I\omega = \left(mr^2\right)\left(\dfrac{2\pi}{\tau}\right)$

Therefore, $H = mr^2\,\dfrac{2\pi}{\tau}$

$$= 0.5\,\text{kg}\,(2\text{m})^2\left(\dfrac{2\text{A}}{10}\,\dfrac{\text{rad}}{\text{sec}}\right)$$

$$= 1.256\ \text{kg.m}^2\text{sec}^{-1}$$

Answer: 1.256 kg.m²/sec

5.5 TORQUE OR MOMENT OF A FORCE

Moment of a force means the action of force which creates rotation of a body. Torque means the action of force which causes twisting of a body. However, both moment of a force and torque are mathematically the same. That is torque or moment of force about a chosen axis is the product of the force and its moment arm. A rigid body can rotate about a point. For example, a photograph can rotate about the point of contact between the nail and the thread from which the photograph is hung; the wheel of a vehicle can rotate about its axis. The magnitude of the moment of force or torque is measured by the product of the magnitude of the force and the perpendicular distance of the action of force from the axis of rotation.

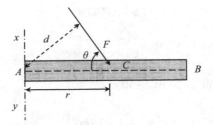

FIGURE 5.5 A Thin Sheet

Let us consider a thin sheet AB, fixed in a horizontal position in such a way that it can rotate about the point A and about the perpendicular axis xy, as shown in Figure 5.5. If the sheet is rotated by applying a force on any point, C, then the produced torque (M) is:

$$\vec{M} = \vec{d} \times \vec{F} = rF \sin \theta$$

Thus, for an object moving about an axis, the cross-product of the position vector of a point on the object where the force is acting, and the applied force is called the torque.

5.6 RELATIONSHIP BETWEEN TORQUE AND ANGULAR ACCELERATION

From the definition of torque, Torque = Force × Perpendicular Distance

Let the mass of an object be m, and the angular acceleration of the object be α, and the angular velocity ω. Then:

Angular acceleration, $\alpha = \dfrac{d\omega}{dt}$

Linear acceleration, $a = \alpha r = r\dfrac{d\omega}{dt}$

Force applied, $F = ma = m\left(r\dfrac{d\omega}{dt} \right) = mr\dfrac{d\omega}{dt}$

Torque, $M = Fr = mr^2 \dfrac{d\omega}{dt}$

$$= I\dfrac{d\omega}{dt}$$

$$= I\alpha$$

In vector, it can be written as, $\vec{M} = I\vec{\alpha}$

Example 5.9

A wheel has a mass of 5 kg and radius of gyration of 25 cm. Calculate its mass moment of inertia. In order to produce angular acceleration of 4 rad/sec² in the wheel calculate the magnitude of torque to be applied.

SOLUTION

Mass, $m = 5$ kg
Radius of gyration, $r = 25$ cm
Angular acceleration, $\alpha = 4$ rad/sec²

Mass moment of inertia, $I = mr^2 = 5\text{kg}(0.25\text{m})^2 = 0.3125 \text{ kg.m}^2$

Torque, $M = I\alpha = 0.3125 \text{ kg.m}^2 \left(4\dfrac{\text{rad}}{\text{sec}^2}\right) = 1.25\text{N.m}$

Answers: 0.3125 kg.m², 1.25 N.m

5.7   KINETIC ENERGY

Kinetic energy is the ability of a rigid body to do work while in motion. Depending on the type of motion, kinetic energy can be expressed as:

For Translation. When a rigid body of mass m is subjected to either rectilinear or curvilinear translation (i.e. no rotation), the kinetic energy (T) of the body is:

$$T = \frac{1}{2}mv^2$$

where v is the translation velocity at the instant considered.

For Rotation about a Fixed Axis. When a rigid body of mass m is subjected to rotation about a fixed axis, the kinetic energy (T) of the body is:

$$T = \frac{1}{2}mv^2 = \frac{1}{2}m(\omega r)^2 = \frac{1}{2}(mr^2)(\omega)^2 = \frac{1}{2}I_o\omega^2$$

where I_o is the mass moment of inertia of the body about the axis perpendicular to the plane of motion and passing through the fixed point of rotation.

For General Plane Motion. When a rigid body of mass m is subjected to general plane motion (both rotation about a fixed axis and translation), the kinetic energy (T) of the body is:

$$T = \frac{1}{2}I_o\omega^2 + \frac{1}{2}mv^2$$

where I_o is the mass moment of inertia of the body about the axis perpendicular to the plane of motion and passing through the fixed point of rotation.

Example 5.10

A 1,000-kg block is to be rolled over a frictionless surface at a velocity of 0.1 m/sec over a 0.3-m radius frictionless pulley applying a tension, as shown in Figure 5.6. Determine the kinetic energy of the 1,000-kg block and the 12-kg pulley.

SOLUTION

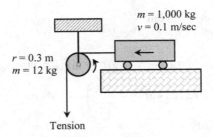

FIGURE 5.6   A Pulley System for Example 5.10

1,000-KG BLOCK

Mass, $m = 1,000$ kg
Velocity, $v = 0.1$ m/sec

It is subjected to rectilinear translation (i.e. no rotation) with velocity (v) of 0.1 m/s. The kinetic energy (T) of the block is:

$$T = \frac{1}{2}mv^2 = \frac{1}{2}(1,000 \text{ kg})\left(0.1 \frac{\text{m}}{\text{sec}}\right)^2 = 5 \text{ J}$$

12-KG PULLEY

Mass, $m = 12$ kg
Radius, $r = 0.3$ m

It is subjected to rotation about a fixed axis. The kinetic energy (T) of the pulley is:

$$T = \frac{1}{2}I_o\omega^2$$

Now, $I_o = \frac{1}{2}mr^2 = \frac{1}{2}(12\text{kg})(0.3 \text{ m})^2 = 0.54 \text{ kg.m}^2$

$$v = \omega r$$

$$\omega = \frac{v}{r} = \frac{0.1 \frac{\text{m}}{\text{sec}}}{0.3 \text{ m}} = 0.33 \frac{\text{rad}}{\text{sec}} \text{ [linear velocity of the block and the pulley}$$

are equal as there is no slippage]

Therefore, $T = \frac{1}{2}I_o\omega^2 = \frac{1}{2}(0.54\,\text{kg.m}^2)\left(0.33\frac{\text{rad}}{\text{sec}}\right)^2 = 0.03\text{ J}$

Answers: 5 J, 0.03 J

Example 5.11

A 1,000-kg cylindrical block is to be rolled over a frictionless surface over a 0.3-m radius frictionless pulley using a 500-kg block, as shown in Figure 5.7. Determine the kinetic energy of the 500-kg block, 12-kg pulley, and 1,000-kg block.

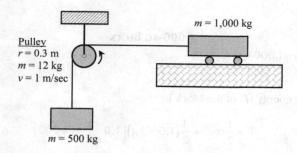

Pulley
$r = 0.3$ m
$m = 12$ kg
$v = 1$ m/sec

$m = 1,000$ kg

$m = 500$ kg

FIGURE 5.7 A Pulley System for Example 5.11

SOLUTION

500-KG BLOCK

Mass, $m = 500$ kg
Velocity, $v = 1.0$ m/sec

It is subjected to rectilinear translation. The linear velocity of the pulley and the linear velocity of the block are equal. The kinetic energy (T) of the block is:

$$T = \frac{1}{2}mv^2 = \frac{1}{2}(500\,\text{kg})\left(1.0\frac{\text{m}}{\text{sec}}\right)^2 = 250\text{ J}$$

12-KG PULLEY

Radius, $r = 0.3$ m
Mass, $m = 12$ kg
Velocity, $v = 1$ m/sec

It is subjected to rotation about a fixed axis. The kinetic energy (T) of the pulley is:

$$T = \frac{1}{2}I_o\omega^2$$

Now, $I_o = \dfrac{1}{2}mr^2 = \dfrac{1}{2}(12\,\text{kg})(0.3\ \text{m})^2 = 0.54\ \text{kg.m}^2$

$$v = \omega r$$

$$\omega = \frac{v}{r} = \frac{1.0\ \dfrac{\text{m}}{\text{sec}}}{0.3\ \text{m}} = 3.33\ \frac{\text{rad}}{\text{sec}}$$

Therefore, $T = \dfrac{1}{2}I_o\omega^2 = \dfrac{1}{2}\left(0.54\ \text{kg.m}^2\right)\left(3.33\dfrac{\text{rad}}{\text{sec}}\right)^2 = 3\ \text{J}$

1,000-KG BLOCK

Mass, $m = 1,000$ kg
Velocity, $v = 1.0$ m/sec

The kinetic energy (T) of the block is:

$$T = \frac{1}{2}mv^2 = \frac{1}{2}(1,000\,\text{kg})\left(1.0\frac{\text{m}}{\text{sec}}\right)^2 = 500\,\text{J}$$

Answers: 250 J, 3 J, 500 J

5.8 WORK, POWER, AND EFFICIENCY

5.8.1 WORK

The definition of work has been described in Chapter 4. Very briefly, the product of applied force, and the resulting displacement along the force, is known as work. If the displacement occurs along the direction making an angle θ with the applied force, then work is equal to the product of the force and the component of displacement along the direction of force. Mathematically, it can be shown as follows:

$$W = Fs\cos\theta$$

In vector form, the work can be presented as:

$$W = \vec{F} \cdot \vec{s}$$

$U_{1\rightarrow 2}$ is often used to denote work done on a particle when it moves from position 1 to position 2.

5.8.2 WORK DONE IN SPRING

Let one end of a horizontal ideal spring S be fastened to a wall and an object of mass m be attached to the other end, as shown in Figure 5.8. Now, if pulling the

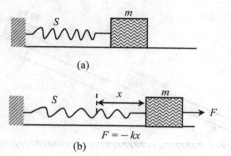

FIGURE 5.8 A Spring System

object, spring S is strained along its length, then a restoring force will develop in the spring which will act against the applied force. According to Hooke's Law, within the elastic limit, the magnitude of the restoring force will be equal to the applied force.

Let x be the extension of length of the spring along the horizontal direction due to the application of horizontal force, F. Due to this action, a restoring force $-kx$ will be developed. This is because:

$$F \propto -x$$

$$F = -kx$$

where k is called spring constant or the spring stiffness. This value represents the amount of force required to cause unit deformation of the spring. In order to expand the spring, the equal amount of external force ($F = kx$) needs to be applied. In expanding the spring from position x_1 to position x_2, work done is given by:

$$W = \int_{x_1}^{x_2} \vec{F} d\vec{x} = \int_{x_1}^{x_2} (kx) dx = k \left[\frac{x^2}{2} \right]_{x_1}^{x_2} = \frac{1}{2} k \left[x_2^2 - x_1^2 \right]$$

This work is positive and remains stored as potential energy in the spring. If the initial position of the spring is at the neutral position, $x_1 = 0$, and the final position is x, then from the above equation, the work done is:

$$W = \frac{1}{2} kx^2$$

Example 5.12

A 20 kg mass, released from rest, slides 6 m down a frictionless plane inclined at an angle of 30° with the horizon and strikes a spring of unknown spring constant, as shown in Figure 5.9. Assume that the spring is ideal, the mass of the spring is negligible, and mechanical energy is conserved.

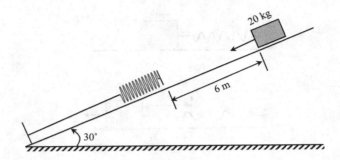

FIGURE 5.9   A Frictionless Inclined Plane for Example 5.12

(a) Determine the speed of the block just before it hits the spring.
(b) Determine the spring constant given that the distance the spring compresses along the incline is 0.3 m when the block comes to rest.

SOLUTION

(a) Speed of the block just before it hits the spring

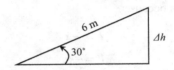

FIGURE 5.10   For Example 5.13

From Figure 5.10, $\Delta h = 6 \sin 30$ m

When the block slides, the potential energy decreases and the kinetic energy increases.

Decrease in Potential Energy = Increase in Kinetic Energy

$mg\Delta h = \frac{1}{2}mv^2$
$g\Delta h = \frac{1}{2}v^2$
$(9.81 \text{ m/sec}^2)(6 \sin 30°) = \frac{1}{2}v^2$
$v = 7.67$ m/sec

(b) Spring constant
Let the spring be compressed by Δx. The total vertical fall of the block,

$h' = \Delta h + \Delta x \sin\theta$
$h' = 6 \text{ m} \sin \theta + \Delta x \sin\theta$
Decrease in Potential Energy = Work Done on the Spring

$mgh' = \frac{1}{2} k\Delta x^2$
$mg(6 \text{ m} \sin\theta + \Delta x \sin\theta) = \frac{1}{2} k\Delta x^2$
$(20 \text{ kg})(9.81 \text{ m/sec}^2)(6 \text{ m} \sin 30 + 0.3 \text{ m} \sin 30) = \frac{1}{2} k (0.3 \text{ m})^2$
Therefore, $k = 13{,}734$ N/m

Answers: (a) 7.67 m/sec
b) 13,734 N/m

5.8.3 WORK DONE ON A STRAINED BODY

Now, we will discuss the work done for extension of a wire or rod, i.e., work done
in case of longitudinal strain. If external force is applied along the length of a
wire, then longitudinal stress is applied in the wire and consequently the length
of the wire increases. Therefore, there occurs a displacement of the point of application of the force, i.e., work is done. Within the proportional limit, the work is
stored as potential energy in the wire. Suppose L is the initial length of a wire of
uniform cross-section and A is the cross-sectional area, as shown in Figure 5.11.
Due to the application of force F, the length of the wire increases by ΔL.

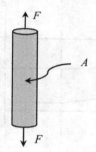

FIGURE 5.11   A Wire of Uniform Cross-Section

Then, the applied stress, $\sigma = \dfrac{F}{A}$

The produced strain, $\varepsilon = \dfrac{\Delta L}{L}$

From Hooke's Law, $E = \dfrac{\sigma}{\varepsilon} = \dfrac{\dfrac{F}{A}}{\dfrac{\Delta L}{L}} = \dfrac{FL}{A\Delta L}$

From the above equation, $F = \dfrac{EA\Delta L}{L}$

During the increase in length, ΔL the average force, $F_{avg} = \dfrac{F}{2} = \dfrac{EA\Delta L}{2L}$

Work, W = Average Force × Displacement = $\dfrac{EA\Delta L}{2L}(\Delta L) = \dfrac{1}{2}\dfrac{EA(\Delta L)^2}{L}$

Therefore, $W = \dfrac{1}{2}\left(\dfrac{EA\Delta L}{L}\right)\Delta L = \dfrac{1}{2}F(\Delta L) = F\left(\dfrac{\Delta L}{2}\right)$

= Force × Avg. Displacement

5.8.4 WORK DONE BY COUPLE MOMENT

If a couple moment works on a body, the body undergoes rotation. For example, in Figure 5.12, the body is experiencing a couple moment of $M = Fr$ by which the body will rotate. If the body rotates θ_1 to θ_2 in a certain amount of time, the work done by this couple moment can be expressed as follows:

$$U_{1 \rightarrow 2} = \int_{\theta_1}^{\theta_2} M \, d\theta$$

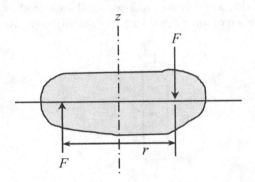

FIGURE 5.12   A Body Experiencing a Couple Moment

5.8.5 POWER

Sometimes, it is important to know how quickly a work can be done rather than the amount of work. For example, a pump capable of drawing 1 million gallons of water during its lifetime may be worthier than a pump capable of drawing 2 million gallons. This is possible if the first pump can do it in 5 years and the second one takes 25 years. Therefore, rate of work is sometimes important when you consider how busy life is today. Power is the time rate of doing work of an agent, and it is measured by work done in unit time. It can be shown as:

$$\text{Power} = \frac{\text{Work}}{\text{Time}}$$

Its unit is Joule/sec or Watt. Another, larger, unit is also used called kilowatt (kW). Another unit, horsepower (hp) is also often used, 1 hp = 746 W.

Sometimes, power is also related to velocity, as follows:

$$\text{Power} = \frac{\text{Work}}{\text{Time}} = \frac{W}{t} = \frac{Fs}{t} = F\left(\frac{s}{t}\right) = Fv$$

In vertical displacement, power can be written as:

$$\text{Power, } P = \frac{W}{t} = \frac{mgh}{t}$$

If the displacement of a body, instead of being along the force, is along a direction making an angle θ with the applied force, then:

$$\text{Power} = Fv\cos\theta = \vec{F}.\vec{v}$$

In the case of rotational motion, we know,

$$\text{Work} = \text{Torque}\left(M\right) \times \text{Angular Displacement}\left(\theta\right)$$

$$\text{Power} = \frac{\text{Work}}{\text{Time}} = \frac{M\theta}{t}$$

Example 5.13

Calculate how much energy per sec (i.e., power) is to be used to lift a stone of mass of 300 kg on top of a roof at a speed of 0.1 m/sec by use of a crane.

SOLUTION

Mass, $m = 300$ kg
Velocity, $v = 0.1$ m/sec

Power, $P = ?$

$$\text{Power, } P = Fv = \left(mg\right)v$$

$$= 300\,\text{kg} \times 9.81\text{m/sec}^2 \times 0.1\text{m/sec}$$

$$= 294\,\text{J/sec}$$

Answer: 294 W

Example 5.14

A ladder of 7.46 m length is in an inclined position at an angle of 60° with the wall of the building. A person weighing 60 kg climbs up to the roof in 30 sec using the ladder taking a load of 15 kg on his head. Calculate the applied power.

SOLUTION

Mass, $m = (60 + 15)$ kg
Height of travel, $h = 7.46 \cos60$ m (shown in Figure 5.13)

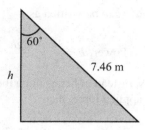

FIGURE 5.13   Ladder in a Building for Example 5.14

Power, $P = ?$

$$\text{Power, } P = \frac{mgh}{t}$$

$$= \frac{(60\,\text{kg} + 15\,\text{kg})\left(9.81\,\dfrac{m}{\text{sec}^2}\right)(7.46\cos60\ \text{m})}{30\ \text{sec}} = 91.4\,\text{W}$$

Answer: 91.4 W

.

Example 5.15

A load of 270 kg mass is pulled up by a crane with a velocity of 0.1 m/sec. Calculate the power of the crane.

SOLUTION

Mass, $m = 270$ kg
Velocity, $v = 0.1$ m/sec

Power, $P = ?$

$$\text{Power, } P = Fv = (mg)v$$

$$= (270\ \text{kg})\left(9.81\,\frac{m}{\text{sec}^2}\right)\left(0.1\,\frac{m}{\text{sec}}\right) = 265\ \text{W}$$

Answer: 265 W

Example 5.16

The depth and diameter of a well full of water are 12 m and 1.8 m, respectively. A pump can empty the well in 24 min. Calculate the power of the pump.

SOLUTION

Mass, $m = $ volume × density

$$= (\pi r^2 h)\rho$$

$$= (\pi)(0.9\text{m})^2 (12\text{m})\left(1,000\,\frac{\text{kg}}{\text{m}^3}\right)$$

$$= 30,536\,\text{kg}$$

Power, $P = \dfrac{mgh}{t}$

$$= \frac{(30,536\,\text{kg})\left(9.81\,\dfrac{\text{m}}{\text{sec}^2}\right)\left(\dfrac{12}{2}\,\text{m}\right)}{24(60)\,\text{sec}}$$

$$= 1,246\,\text{W}$$

$$= \frac{1,246\,\text{W}}{746}$$

$$= 1.67\,\text{hp}$$

Answer: 1.67 hp

Note: The pump draws all water from a 12-m deep well, then the height of drawing water is 12 m divided by 2, i.e., 6 m. The reason for this is that at the beginning, the water surface is at the depth of 0 and at the end the water is at 12 m. The average depth of the well is then the average depth of drawing water.

5.8.6 EFFICIENCY

When we get work from a machine or a body, that work is smaller than the energy needed to supply for that machine or body to do the work. It is not only true for machines, but is also true in our practical lives as well where a part of the work is utilized and the rest of the supplied energy is lost. In the case of an engine, this dissipation of energy is used as friction, to warm up the engine, etc. This dissipation of energy cannot be stopped completely, but by applying different technologies this dissipation can be reduced. In this case, equation of energy is equivalent to supplied energy = effective energy + dissipated energy. The efficiency of a machine is defined as the ratio of the output energy to the input or supplied energy. It is commonly denoted as η.

$$\eta = \frac{\text{Output Energy}}{\text{Input Energy}}$$

Let E_1 be the energy supplied and E_2 be the dissipation of energy. Then, the output is $E_1 - E_2$.

$$\eta = \frac{E_1 - E_2}{E_1} = \left(1 - \frac{E_2}{E_1}\right) \times 100\%$$

Example 5.17

A motor of power 3,430 W pumps water to a height of 7.20 m from a well. If the efficiency of the motor is 90%, how much water can it pump in a minute?

SOLUTION

Power, $P = 3,430$ W
Depth of well, $h = 7.20$ m
Working time, $t = 60$ sec
Efficiency, $\eta = 0.90$
Mass of water, $m = ?$

Let the mass be m kg. The efficiency of the motor is 90%, meaning the effective power is $= 0.9(3,430 \text{ W}) = 3,087$ W.

$$\text{Power, } P = \frac{W}{t} = \frac{mgh}{t}$$

$$3,087 W = \frac{m\left(9.81\dfrac{m}{sec^2}\right)(7.2m)}{60 \, sec}$$

Therefore, $m = 2,622$ kg
Answer: 2,622 kg

Example 5.18

An engine pumps 1,000 kg of water per minute from a well of depth 10 m. If 40% power of the engine is lost, determine the horsepower of the engine.

SOLUTION

Mass of water, $m = 1,000$ kg
Working time, $t = 60$ sec
Depth of well, $h = 10$ m

Power, $P = ?$
Given 40% power is lost, the efficiency is $100 - 40 = 60\%$

$$0.60(P) = \frac{mgh}{t}$$

$$0.60(P) = \frac{(1,000\,\text{kg})\left(9.81\dfrac{m}{\sec^2}\right)(10\,\text{m})}{60\,\sec}$$

Therefore, $P = \dfrac{(1,000\ \text{kg})\left(9.81\dfrac{m}{\sec^2}\right)(10\,\text{m})}{60\,\sec}\left(\dfrac{1}{0.60}\right)$

$$= 2,722\ W = \frac{2,722\ W}{746\ \dfrac{W}{hp}}$$

$$= 3.65\,hp$$

Answer: 3.65 hp

5.9   PRINCIPAL OF WORK AND ENERGY

The particle's initial kinetic energy plus the work done by all the forces acting on the particle as it moves from its initial to its final position equals the final kinetic energy of the particle. That means:

$$T_1 + U_{1\to 2} = T_2$$

where T_1 and T_2 are the initial and the final kinetic energy, respectively; $U_{1\to 2}$ is the work done by the body. This equation states that the initial translation and rotational kinetic energy plus the work done on the body by external forces and moments equal the final translation and rotational kinetic energy.

Example 5.19

A 0.3-m radius circular disk with a mass of 2 kg is subjected to the couple moment of $M = 1.5\theta + 2$ where θ is in radians and M is in N.m, as shown in Figure 5.14. If the body starts from rest, determine its angular velocity when it has made 5 revolutions.

SOLUTION

Radius, $r = 0.3$ m
Mass, $m = 2$ kg

Mass moment of inertia, $I_o = \dfrac{1}{2}mr^2$

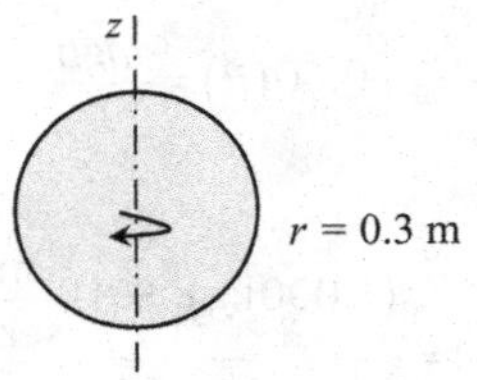

FIGURE 5.14   The 0.6-m Circular Disk for Example 5.19

$$= \frac{1}{2}(2 \text{ kg})(0.3 \text{ m})^2$$

$$= 0.09 \text{ kg.m}^2$$

Kinetic energy, $T = \frac{1}{2}I_o\omega^2$

Initial kinetic energy, $T_1 = \frac{1}{2}I_o\omega_1^2 = \frac{1}{2}I_o(0)^2 = 0$

Final kinetic energy, $T_2 = \frac{1}{2}I_o\omega_2^2 = \frac{1}{2}(0.09)(\omega_2)^2 = 0.045(\omega_2)^2$

Due to the work done on the body, the kinetic energy of the disk changes. From the principal of work and energy:

$$T_1 + U_{1\to2} = T_2$$

$$T_1 + \int_{\theta_1}^{\theta_2} M\, d\theta = T_2$$

$$0 + \int_{0}^{10\pi} (1.5\theta + 2)\, d\theta = 0.045\omega_2^2$$

$$\left[\frac{1.5\theta^2}{2} + 2\theta \right]_0^{10\pi} = 0.045\omega_2^2$$

$$0.045\omega_2^2 = \left[\frac{1.5\theta^2}{2} + 2\theta \right]_0^{10\pi}$$

$$0.045\omega_2^2 = \left[\frac{1.5(10\pi)^2}{2} + 2(10\pi) \right] - \left[\frac{1.5(0)^2}{2} + 2(0) \right]$$

$$0.045\omega_2^2 = 803.1$$

Therefore, $\omega_2 = 133.6 \dfrac{\text{rad}}{\text{sec}}$

Answer: 133.6 rad/sec

5.10 IMPULSE AND MOMENTUM

Linear Impulse: Impulse (or linear impulse) is defined as the product of the force and its time of action. It can be expressed as follows:

$$\vec{J} = \vec{F}t$$

Linear Momentum: Linear momentum is the product of the mass of the rigid body and its velocity. It is a vector quantity and can be written as:

$$\vec{P} = m\vec{v}$$

where m is the mass and v is the velocity of the rigid body.

Angular Impulse: If a force of F is applied to a body from t_1 to t_2, with a perpendicular distance of r from the reference axis, then the angular impulse of the above system can be expressed as:

$$\int_{t_1}^{t_2} M_o \, dt = \int_{t_1}^{t_2} \left(\vec{r} \times \vec{F} \right) dt$$

Angular Momentum: Angular momentum of a particle is defined as the moment of a particle's linear momentum about an axis. Thus, sometimes, it is referred to as the moment of momentum. Let a particle A of mass m be moving in a curvilinear path in the xy plane, as shown in Figure 5.15. The perpendicular distance is r from point A to the perpendicular axis of the xy plane passing through O as shown below.

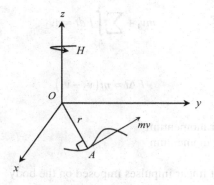

FIGURE 5.15 A Curvilinear Path in the xy Plane

At the instant shown, the velocity of the particle is v. Then, the linear momentum of the particle is mv. Thus, the angular momentum can be presented as:

$$H = (mv)r = m(\omega r)r = (mr^2)\omega = I\omega$$

Thus, angular momentum is defined as the product of the moment of inertia of the body about the axis passing through the mass center and the angular velocity of the rigid body. It can be written as:

$$H = I\omega$$

where I is the mass moment of inertia about the axis passing through the mass center and ω is the angular velocity. For rotation about a fixed axis, angular momentum can be written as:

$$H_o = I_o\omega$$

where I_o is the mass moment of inertia about the axis perpendicular to the fixed rotation point and passing through it. More of these definitions can be found in the previous chapter.

5.11   CONSERVATION OF IMPULSE AND MOMENTUM

Conservation of linear momentum states that the sum of all impulses by the external forces in a period equals the change of the linear momentum of the body during that time frame. It can be written as:

$$\sum \int_{t_1}^{t_2} F\,dt = mv_2 - mv_1$$

or,

$$mv_1 + \sum \int_{t_1}^{t_2} F\,dt = mv_2$$

or,

$$F\Delta t = m(v_2 - v_1)$$

where
   mv_1 = Initial linear momentum
   mv_2 = Final linear momentum

$$\sum \int_{t_1}^{t_2} F\,dt = \text{sum of linear impulses imposed on the body}$$

F = force applied

Δt = time of impulse

Similarly, conservation of angular momentum can be written as:

$$\left(H_1\right)_z + \sum \int_{t_1}^{t_2} M_z dt = \left(H_2\right)_z$$

where

$\left(H_1\right)_z$ = Initial angular momentum

$\left(H_2\right)_z$ = Final angular momentum

$\sum \int_{t_1}^{t_2} M_z dt$ = sum of the angular impulses created on the body

Example 5.20

The Class 9 Truck that is commonly seen on highways has a total weight of 72 kip. The maximum speed of the road is 60 mph. If the truck hits a bridge and stops in 0.5 sec, calculate the collision load that the bridge will react to the truck.

SOLUTION

$$\text{Mass}, m = \frac{W}{g} = \frac{72 \text{kip}}{32.2 \dfrac{\text{ft}}{\text{sec}^2}} = \frac{72,000 \text{ lb}}{32.2 \dfrac{\text{ft}}{\text{sec}^2}} = 2,236 \text{ slug}$$

$$\text{Initial speed}, v_1 = 60 \frac{\text{mile}}{\text{hour}} = \frac{60 \text{ mile} \left(5,280 \dfrac{\text{ft}}{\text{mile}}\right)}{(1 \text{hour}) 3,600 \dfrac{\text{sec}}{\text{hour}}} = 88 \frac{\text{ft}}{\text{sec}}$$

Final speed, $v_2 = 0$

Time lapse, $\Delta t = 0.5$ sec

$$F\Delta t = m\left(v_2 - v_1\right)$$

$$F\left(0.5 \text{sec}\right) = \left(2,236 \text{slug}\right)\left(0 - 88 \frac{\text{ft}}{\text{sec}}\right)$$

Therefore, $F = 393,536$ lb ≈ 394 kip

Answer: 394 kip

Example 5.21

A disk has a weight of 20 lb, radius of 0.3 ft, and is pinned at its center, as shown in Figure 5.16. If a vertical force of 50 lb is applied to the cord wrapped around its outer rim, determine the angular velocity of the disk in 5 sec starting from rest. Ignore the mass of the cord.

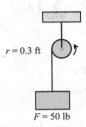

$r = 0.3$ ft

$F = 50$ lb

FIGURE 5.16   A Circular Disk for Example 5.21

SOLUTION

Weight of disk, $W = 20$ lb

Mass of disk, $m = \dfrac{W}{g} = \dfrac{20 \text{ lb}}{32.2 \dfrac{\text{ft}}{\text{sec}^2}}$

Applied force, $F = 50$ lb
Travel time, $t = 5$ sec

Due to the application of angular impulse, the angular momentum of the body changes. From the conservation of angular momentum:

$$(H_1)_o + \sum \int_{t_1}^{t_2} M_o dt = (H_2)_o$$

$$I_o \omega_1 + \sum \int_{t_1}^{t_2} M_o dt = I_o \omega_2$$

$$I_o \omega_1 + \sum \int_{t_1}^{t_2} (rF) dt = I_o \omega_2$$

$$I_o(0) + 50 \text{ lb}(0.3 \text{ ft})(5 \text{ sec}) = \left[\frac{1}{2} \left(\frac{20 \text{ lb}}{32.2 \dfrac{\text{ft}}{\text{sec}^2}} \right)(0.3 \text{ ft})^2 \right] \omega_2$$

Therefore, $\omega_2 = 2,683 \dfrac{\text{rad}}{\text{sec}}$

Answer: 2,683 rad/sec

FUNDAMENTALS OF ENGINEERING (FE)
EXAM STYLE QUESTIONS

FE Problem 5.1

A large sphere rolls without slipping across a horizontal surface, as shown in Figure 5.17. The sphere has a constant translational speed of 10 m/sec, a mass (m) of 25 kg, and a radius of 0.2 m. The moment of inertia of the sphere about its center of mass is $I = 2/5\ mr^2$. The kinetic energy (J) of the sphere as it rolls along the horizontal surface is most nearly:

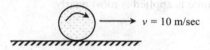

$v = 10$ m/sec

FIGURE 5.17   Rolling of a Sphere for FE Problem 5.1

   A. 1,250
   B. 500
   C. 1,750
   D. 2,250

FE Problem 5.2

A large sphere rolls without slipping across a horizontal surface, as shown in Figure 5.18. The sphere has a constant translational speed of 10 m/sec, a mass (m) of 25 kg, and a radius of 0.2 m. The moment of inertia of the sphere about its center of mass is $I = 2/5\ mr^2$. The sphere approaches a 30° incline of height 3 m and rolls up the incline without slipping. The kinetic energy (J) of the sphere considering both translation and rotation can be calculated using $7/10\ mv^2$. The magnitude of the sphere's velocity (m/sec) just as it leaves the top of the incline is most nearly:

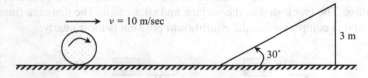

$v = 10$ m/sec

3 m

30°

FIGURE 5.18   Rolling of a Sphere for FE Problem 5.2

   A. 3.4
   B. 5.2
   C. 7.6
   D. 8.4

FE Problem 5.3

If a spring is elongated by 15 cm, its potential energy is 60 J. The amount of force (kN) required to elongate this spring by 30 cm is most nearly:

A. 1,600
B. 1280
C. 42
D. 1.6

FE Problem 5.4

A 2-kg object initially moving with a constant velocity is subjected to a force of magnitude F in the direction of motion. A graph of F as a function of time t is shown in Figure 5.19. The increase, if any, in the velocity (m/sec) of the object during the time the force is applied is most nearly:

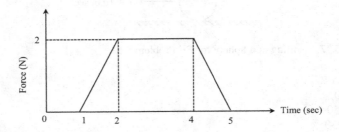

FIGURE 5.19   A Graph of F-t for FE Problem 5.4

A. 0
B. 3.0
C. 4.0
D. 6.0

FE Problem 5.5

A 2-kg block is dropped from a height of 0.45 m above an elastic ground, as shown in Figure 5.20. The ground has an elastic constant of 200 N/m and negligible mass. The block strikes the surface and sticks to it. The distance (mm) that the ground is compressed at the equilibrium position is most nearly:

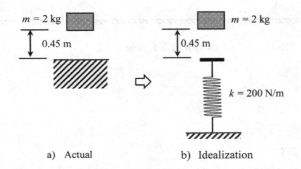

a) Actual   b) Idealization

FIGURE 5.20   Elastic Deformation of Ground for FE Problem 5.5

A. 0.098
B. 98
C. 45
D. 129

FE Problem 5.6

A ball is held motionless at a height of h_o above a hard floor and released. Assuming that the collision with the floor is elastic, which one of the following graphs best shows the relationship between the total energy, E, of the ball and its height, h, above the surface?

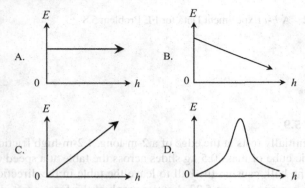

FE Problem 5.7

Three objects can only move along a straight, level path. On the graphs shown in Figure 5.21 the position, s, of each of the objects is plotted as a function of time, t. The magnitude of the momentum of the object is increasing in which of the cases?

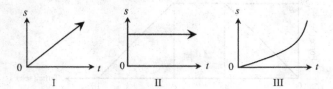

FIGURE 5.21 Distances Traveled by Three Objects for FE Problem 5.7

A. II only
B. III only
C. I and II only
D. I and III only

FE Problem 5.8

A student obtains data on the magnitude of force applied to an object as a function of time and displays the data on the graph shown in Figure 5.22. The increase in the momentum (N.sec) of the object from $t=0$ to $t=4$ sec is most nearly:

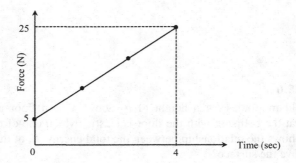

FIGURE 5.22 A *F-t* Experiment Data for FE Problem 5.8

A. 25
B. 50
C. 60
D. 100

FE Problem 5.9

A 5-kg ball initially rests at the edge of a 2-m-long, 1.2-m-high frictionless table. A hard plastic cube of mass 0.5 kg slides across the table at a speed of 26 m/sec and strikes the ball, causing the ball to leave the table in the direction in which the cube was moving. Figure 5.23 shows a graph of the force exerted on the ball by the cube as a function of time. The total impulse (N.sec) given to the ball is most nearly:

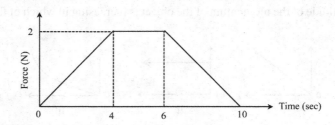

FIGURE 5.23 A *F-t* Data for FE Problem 5.9

A. 20
B. 15
C. 12
D. 10

FE Problem 5.10

A 2.0-kg frictionless cart is moving at a constant speed of 3.0 m/sec to the right on a horizontal surface, as shown in Figure 5.24, when it collides with a second

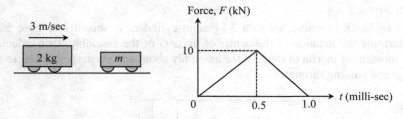

FIGURE 5.24   A *F-t* Data of a Frictionless Cart for FE Problem 5.10

cart of undetermined mass *m* that is initially at rest. The force *F* of the collision is a function of time *t*, where $t=0$ is the instant of initial contact. As a result of the collision, the second cart acquires a speed of 1.6 m/sec to the right. Assume that friction is negligible before, during, and after the collision. The magnitude (m/sec) of the velocity of the 2.0-kg cart after the collision is most nearly:

A. 0.5
B. 1.0
C. 1.5
D. 2.0

Practice Problems

Problem 5.1

The mass moment of inertia of a body with respect to an axis is 100 kg.m². Determine the radius of gyration of the body with respect to that axis if the weight of the body is 29.4 N.

Problem 5.2

The pendulum consists of a 3-kg slender rod and a 5-kg thin plate, as shown in Figure 5.25. Determine the location of the center of mass G of the pendulum; then calculate the moment of inertia of mass of the pendulum about an axis perpendicular to the page and passing through the hanging point.

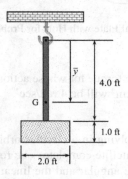

FIGURE 5.25   The Pendulum for Problem 5.2

Problem 5.3

A 3-kg block is connected to a 3-kg semi-cylinder, as shown in Figure 5.26. Determine the location of the center of mass G of the assembly; then calculate the moment of inertia of mass of the assembly about an axis perpendicular to the page and passing through G.

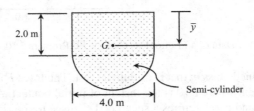

FIGURE 5.26   The Block for Problem 5.3

Problem 5.4

A thin metal plate has the following dimensions with a 1.5-ft diameter hole at its center, as shown in Figure 5.27. The metal weighs 100 pound per square feet (psf).

(a) Calculate the moment of inertia of mass of the plate about an axis perpendicular to the page and passing through the centroid.
(b) Calculate the moment of inertia of mass of the plate about an axis perpendicular to the page and passing through the hanging point.

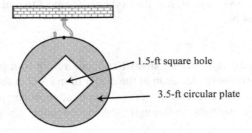

FIGURE 5.27   The Thin Metal Plate with Hole for Problem 5.4

Problem 5.5

Calculate the magnitude of a torque for whose action the angular acceleration for moment of inertia of 250 kg.m² will be 4 rad/sec².

Problem 5.6

An artificial satellite is revolving in a circular orbit at a height of 500 m from the Earth's surface. If the satellite completes one revolution around the Earth in 100 min, then determine the angular and the linear velocity. The Earth's radius is 6,400 km.

Problem 5.7

The mass of a wheel is 5 kg and the radius of gyration with respect to an axis is 0.2 m. Calculate its moment of inertia. What magnitude of torque is to be applied on the wheel to generate angular acceleration of 2 rad/sec²?

Problem 5.8

The diameter of a wheel is 2 m and its mass is 20 kg. Calculate the angular momentum for the wheel at an angular velocity of 1800 rpm.

Problem 5.9

If a piece of stone of mass 0.25 kg fastened at the end of a thread of length 0.75 m completes 90 revolutions per minute, calculate the tension in the string.

Problem 5.10

A rigid body of mas of 5 kg is rotating at a distance of 1.5 m from the axis of rotation with an angular velocity of 5 rad/sec. Calculate its mass moment of inertia and its rotational kinetic energy.

Problem 5.11

The mass moment of inertia of a flywheel is 0.05 kg.m². If its angular velocity increases from 60 rpm to 300 rpm in 8 sec, find the torque acting on the wheel.

Problem 5.12

A body of mass 4 g is rotated by a thread of length 1.5 m in a circular path. The body completes 20 revolutions in 5 sec. Calculate the tension of the thread.

Problem 5.13

Calculate how much water a pump of 1hp can raise to a height of 10 m in 1 min.

Problem 5.14

A person weighing 70 kg can ascend 30 steps, each of which 12 cm in height, of a ladder in 20 sec. Calculate the power of the person.

Problem 5.15

A person weighing 150 kg descends along a ladder of 4 m long taking a load of 50 kg. If the ladder makes an angle of 60° to the wall, calculate the work done by him.

Problem 5.16

The depth and diameter of a well are 7.2 m and 4 m, respectively. A pump can make the well empty in 31.4 min. Calculate the power of the pump.

Problem 5.17

A motor of 6 kW power pumps water to a height of 20 m from a well. If the efficiency of the motor is 82.2%, how much water can be pumped in 1 min?

Problem 5.18

A motor of 80% efficiency controls a crane whose efficiency is 50%. If the motor applies 3.73 kW power to the crane to lift a body of 746 N, calculate the average vertical velocity of the body.

Problem 5.19

To raise the angular velocity of a flywheel from 2π to 6π rad/sec, work of 100 J is done. Calculate the mass moment of inertia of the wheel.

Problem 5.20

The radius vector of a rotating particle is $\vec{r} = 2\vec{i} + 2\vec{j} - \vec{k}$ m, and applied force (N) is $\vec{F} = 6\vec{i} + 3\vec{j} - 3\vec{k}$. Determine the torque both in scalar and vector formats.

Problem 5.21

The 1.0-ft circular disk with a mass of 2 lb is subjected to the couple moment of $M = 0.5\theta + 2$ where θ is in radians and M is in lbf.ft, as shown in Figure 5.28. If the body starts from the rest, determine its angular velocity when it has made 4 revolutions.

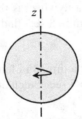

FIGURE 5.28   The Circular Disk for Problem 5.21

Problem 5.22

A disk has a weight of 20 lb, and diameter (d) of 0.6 ft and is pinned at its center, as shown in Figure 5.29. If the 50-lb block descends 5 ft down, determine the velocity of the block if it starts from rest. Ignore the mass of the cord.

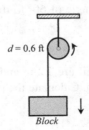

$d = 0.6$ ft

Block

FIGURE 5.29   The Disk for Problem 5.22

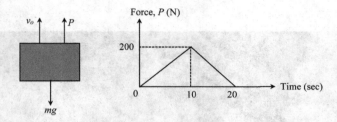

FIGURE 5.30 The Forces on a Block for Problem 5.23

Problem 5.23

The block shown in Figure 5.30 has a mass (m) of 10 kg. At time, $t = 0$ sec, the start velocity is 10 m/sec. The Force-Time ($F\text{-}t$) graph is given. Calculate the velocity of the block at time, $t = 15$ sec. (Hint: apply $\sum \int_{t_1}^{t_2} F dt = mv_2 - mv_1$)

6 Dynamics in Roadways

6.1 GENERAL

Newton's Laws of Motion, discussed in earlier chapters, assume that there is no resistance in motion. For example, while driving a car, there is no air resistance, no frictional force, etc. However, in practical life, it is impossible to have this kind of resistance-free movement. All the roadways have very high frictional resistance, air resistance is present in an ambient environment, the roadway may have longitudinal elevation to drain water, etc. Therefore, to produce acceleration in the vehicle, the engine must apply greater force compared to that of Newton's Second Law (Force = mass × acceleration). The required force (F) to produce a certain acceleration must consider the resistances present in the roadway. Newton's Second Law for practical roadway conditions can be written as:

$$F = ma + \text{Resistance Forces}$$

where

F = the applied force by the vehicle
m = mass of the vehicle
a = acceleration of the vehicle

Now, we will discuss the resistance forces present in a practical roadway, and their computations.

6.2 RESISTANCES

The major resistances that occur in a roadway can be listed as:

(a) Air Resistance
(b) Surface or Rolling Resistance
(c) Grade Resistance

6.2.1 AIR RESISTANCE

Imagine you are driving a light-weight vehicle (say, a car) during a windy day. You can easily feel that the vehicle shakes due to the lateral wind. Similarly, if the wind blows toward your vehicle, it resists the movement of your vehicle; consequently, the speed decreases. This air resistance is very often called aerodynamic resistance. Aerodynamic resistance is a resistive force that can have significant impacts on vehicle performance. At high speeds, where this component

DOI: 10.1201/9781003283959-6

of resistance can become overwhelming, proper vehicle aerodynamic design is essential. Attention to aerodynamic efficiency in design has long been the rule in racing and sports cars. More recently, concerns over fuel efficiency and over-all vehicle performance have resulted in more efficient aerodynamic designs in normal passenger cars, although not necessarily in pickup trucks or sport utility vehicles (SUVs). Aerodynamic resistance originates from a number of sources. The primary source (typically accounting for over 85% of total aerodynamic resistance) is the turbulent flow of air around the vehicle body. This turbulence is a function of the shape of the vehicle, particularly the rear portion, which has been shown to be a major source of air turbulence. To a much lesser extent (on the order of 12% of total aerodynamic resistance), the friction of the air passing over the body of the vehicle also contributes to resistance. Finally, approximately 3% of the total aerodynamic resistance can be attributed to air flow through vehicle components such as radiators and air vents (Mannering and Washburn, 2013). Let us see another example. If you walk or run in opposition to a strong wind, your walking or running speed decreases. While running the decrease in speed is higher. That means the speed of your travel against the wind has a direct impact on the resistance. The higher the speed, the higher the resistance. Therefore, aero-dynamic resistance is higher for an airplane. This resistance force can be calcu-lated as:

$$R_a = \frac{\rho}{2} C_D A_f v^2$$

where:

R_a = Aerodynamic resistance force, lb or N

ρ = Air density in slug/ft^3 or kg/m^3 (commonly, 0.00237 slug/ft^3 or 1.22 kg/m^3)

C_D = Drag Coefficient; common automobiles have coefficient values ranging from 0.25 to 0.55; buses have drag coefficient values ranging from 0.5 to 0.70; motorcycles have drag coefficient values ranging from 0.27 to 1.8.

A_f = Projected area of the vehicle in the direction of travel, ft^2 or m^2

v = Speed of the vehicle (fps or m/sec); wind speed is very often neglected; if wind speed is exactly known, it is better to consider the net speed along the direction of the vehicle with respect to the air. The vehicle speed and component of wind speed in opposition to the vehicle need to be added to determine the net speed. If the component of wind speed is along the vehicle's speed, then it is to be deducted from the vehicle speed.

Air density is a function of both elevation and temperature, as indicated in Table 6.1. As the air becomes denser, total aerodynamic resistance increases.

The drag coefficient (C_D) is a term that implicitly accounts for all three of the aerodynamic resistance sources discussed above. The drag coefficient is mea-sured from empirical data either from wind tunnel experiments or actual field tests in which the vehicle is allowed to decelerate from a known speed with other sources of resistance (rolling and grade) accounted for. The general trend toward

TABLE 6.1

Typical Values of Air Density under Specified Atmospheric Conditions

Altitude (ft)	Temperature (°F)	Pressure (psi)	Air density (slugs/ft³)
0	59	14.7	0.002378
5,000	41.2	12.2	0.002045
10,000	23.4	10.1	0.001755

lower drag coefficients over this period of time reflects the continuing efforts of the automotive industry to improve overall vehicle efficiency by minimizing resistance forces. Also, automobile operating conditions can have a significant effect on drag coefficients. For example, even a small operational change, such as the opening of windows, can increase drag coefficients by 5% or more. More significant operational changes, such as having the top down on a convertible automobile, can increase the drag coefficient by more than 25%. Finally, projected frontal area (approximated as the height of the vehicle multiplied by its width) typically ranges from 10 ft² to 30 ft² (1.0 m² to 2.5 m²) for passenger cars and is also a major factor in determining aerodynamic resistance. Because aerodynamic resistance is proportional to the square of the vehicle's speed, it is clear that such resistance will increase rapidly at higher speeds and vice versa.

Example 6.1

Calculate the force exerted by an 80 mile per hour (mph) wind in front of a standing bus whose projected area is 100 ft². The wind is blowing in the opposite direction to the bus. Assume, the drag coefficient is 0.6, and the air density is 0.00237 slug/ft³.

SOLUTION

Given,

$$\text{Speed, } v = 80 \frac{\text{mile}}{\text{hour}} = \frac{80\,\text{mile}\left(5,280\,\dfrac{\text{ft}}{\text{mile}}\right)}{(1\text{hour})\,3,600\,\dfrac{\text{sec}}{\text{hour}}} = 117\,\frac{\text{ft}}{\text{sec}}$$

$$\text{Or simply, } 80\,\text{mph} = 80\,\text{mph}\left(1.47\,\frac{\text{fps}}{\text{mph}}\right) = 117.6\,\text{fps}$$

Projected area, $A = 100$ ft²

Air density, $\rho = 0.00237$ slug/ft³

Drag coefficient, $C_D = 0.6$

Air resistance, $R_a = \dfrac{\rho}{2} C_D A_f v^2$

$$= \frac{\left(0.00237 \, \dfrac{\text{slug}}{\text{ft}^3}\right)}{2}(0.6)\left(100 \text{ ft}^2\right)\left(117 \, \frac{\text{ft}}{\text{sec}}\right)^2$$

$$= 973 \text{ lb}$$

Answer: 973 lb
Note: Speed here is wind speed with respect to the standing bus.

6.2.2 SURFACE OR ROLLING RESISTANCE

When one surface rolls over another, the roughness between these two surfaces causes friction and tries to hinder the movement of the surfaces. For a vehicle, the friction is generated between the tires and the road surface. The surface resistance can be determined using the friction principle.

The primary source of this resistance is the deformation of the tire as it passes over the roadway surface. The force needed to overcome this deformation accounts for approximately 90% of the total rolling resistance. Depending on the vehicle's weight and the material composition of the roadway surface, the penetration of the tire into the surface and the corresponding surface compression can also be a significant source of rolling resistance.

In considering the sources of rolling resistance, three factors are worthy of note. First, the rigidity of the tire and the roadway surface influence the degree of tire penetration, surface compression, and tire deformation. Hard, smooth, and dry roadway surfaces provide the lowest rolling resistance. Second, tire conditions, including inflation pressure and temperature, can have a substantial impact on rolling resistance. High tire inflation decreases rolling resistance on hard paved surfaces as a result of reduced friction but increases rolling resistance on soft unpaved surfaces due to additional surface penetration. Also, higher tire temperatures make the tire body more flexible, and thus less resistance is encountered during tire deformation. The third and final factor is the vehicle's operating speed, which affects tire deformation. Increasing speed results in additional tire flexing and vibration and thus a higher rolling resistance. Let us consider a block of weight, W, is to be moved applying a force of, F, as shown in Figure 6.1. The frictional coefficient between the surfaces is μ_r. If the normal reaction to the body

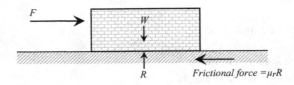

FIGURE 6.1 A Body Is Being Pushed on a Horizontal Surface

is R, then the maximum resistive force of $\mu_r N$ will be generated depending on the value of applied force, F.

If the above block of weight, W, is to be pushed in an inclined plane (θ_g angle with the surface), then the weight will be resolved, as shown in Figure 6.2.

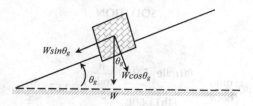

FIGURE 6.2 Weight Components of a Body in an Inclined Surface

Then the frictional force will be:

$$F_f = \mu_r W \cos\theta_g$$

Similarly, in a vehicle, the frictional force or rolling resistance, R_r, is simply the coefficient of friction multiplied by the weight acting normally on the roadway surface and is given as follows:

$$R_r = \mu_r R = \mu_r \left(W \cos\theta_g \right)$$

The longitudinal elevation, θ_g, is often very small and can be ignored. Then, the rolling resistance, R_r, can be determined as:

$$R_r = \mu_r \left(W \cos\theta_g \right) \approx \mu_r \left(W \right)$$

Due to the wide range of factors that determine rolling resistance, a simplifying approximation is used. Studies have shown that overall rolling resistance can be approximated as the product of a friction term (coefficient of rolling resistance) and the weight of the vehicle acting normally to the roadway surface. The coefficient of rolling resistance μ_r for road vehicles operating on paved surfaces is approximated as:

$$\mu_r = 0.01\left(1 + \frac{v}{147} \right) \text{ in US units}$$

$$\mu_r = 0.01\left(1 + \frac{v}{44.73} \right) \text{ in SI units}$$

where
$v=$ speed of the vehicle is fps or m/sec; wind speed is not considered as friction depends on vehicle movement only. If there is a speed variation, then consider the average speed to calculate μ_r.

Example 6.2

The weight of a car is 4,000 lb. If it runs at a speed of 80 mph on a flat or near flat roadway, calculate the surface resistance of this car.

SOLUTION

Given,

$$\text{Speed, } v = 80 \frac{\text{mile}}{\text{hour}} = \frac{80\,\text{mile}\left(5{,}280\,\frac{\text{ft}}{\text{mile}}\right)}{(1\text{h})3{,}600\,\frac{\text{sec}}{\text{h}}} = 117\,\frac{\text{ft}}{\text{sec}}$$

Weight of the car, $W = 4{,}000$ lb

$$\text{Rolling/surface resistance, } R_r = \mu_r (W) = 0.01\left(1 + \frac{v}{147}\right)(W)$$

$$= 0.01\left(1 + \frac{117}{147}\right)(4{,}000\,\text{lb}) = 71.8\,\text{lb}$$

Answer: 71.8 lb

6.2.3 GRADE RESISTANCE

Grade resistance R_g is the gravitational force component parallel to the road surface acting on the vehicle. It can be expressed as:

$$R_g = W \sin\theta_g$$

Let us consider a car is running in an upward grade as shown in Figure 6.3. The weight component along the road surface which forces the car back is $W\sin\theta_g$. This component of weight is considered the grade resistance as it opposes the movement of the car. The longitudinal elevation, θ_g, is often very small and $\sin\theta_g$ can be written as $\tan\theta_g \cong G$. Therefore, the grade resistance can be expressed as:

$$R_g \cong WG$$

$$R_g = W \sin\theta_g$$

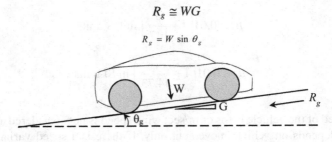

FIGURE 6.3 A Car in an Upward Grade

where:
  G = grade of the roadway; vertical rise per unit horizontal distance (ft/ft or m/m).

It needs to be noted that the grade sign (upward or downward) affects this resistance. The resistance force is negative for a downward grade and vice-versa. This means the downward grade adds value to the applied force and acceleration is faster.

Example 6.3

The weight of a car is 4,000 lb. If it runs at a speed of 80 mph on a 4% upward grade roadway, calculate the grade resistance to this car.

SOLUTION

Given,

$$\text{Speed, } v = 80\,\frac{\text{mile}}{\text{hour}} = \frac{80\,\text{mile}\left(5{,}280\,\frac{\text{ft}}{\text{mile}}\right)}{(1\text{h})3{,}600\,\frac{\text{sec}}{\text{h}}} = 117\,\frac{\text{ft}}{\text{sec}}$$

Weight of the car, W = 4,000 lb
Grade, G = 4% upward

$$\text{Grade resistance, } R_g = WG = (4{,}000\,\text{lb})\left(\frac{4}{100}\right) = 160\,\text{lb}$$

Answer: 160 lb

6.3 EFFECTIVE TRACTIVE FORCE

The tractive effort available to overcome resistance and/or to accelerate the vehicle is determined either by the force generated by the vehicle's engine or by some maximum value that will be a function of the vehicle's weight distribution and the characteristics of the roadway surface–tire interface. Considering the resistances, the required force to produce a certain acceleration, with an additional term to account for the inertia of the vehicle's rotating parts that must be overcome during acceleration, can be given as:

$$F = \gamma_m ma + R_a + R_r + R_g$$

where
  F = applied force by the vehicle, lb or N
  m = mass of the vehicle, slug or kg
  a = acceleration of the vehicle, ft/sec^2 or m/sec^2

R_a = Air resistance, lb or N
R_r = Rolling resistance, lb or N
R_g = Grade resistance, lb or N
γ_m = the mass factor

The mass factor (γ_m) accounts for the inertia of the vehicle's rotating parts that must overcome during acceleration and is approximated as:

$$\gamma_m = 1.04 + 0.0025\varepsilon_o^2$$

Where ε_o = overall gear reduction ratio. The overall gear reduction ratio (ε_o), which includes the gear reductions of the transmission and differential, plays a key role in the determination of tractive effort. By definition, the overall gear reduction ratio refers to the relationship between the revolutions of the engine's crankshaft and the revolutions of the drive wheels. For example, an overall gear reduction ratio of 5 to 1 ($\varepsilon_o = 5$) means that the engine's crankshaft turns five revolutions for every one revolution of the drive wheels. During braking, the value of mass factor (γ_m) is 1.04 for automobiles (Wong 2008). The effective tractive force can be written as:

$$F = \gamma_m ma + R_a + R_r + R_g$$

$$F - R_a - R_r - R_g = \gamma_m ma$$

$$F_{effective} = \gamma_m ma$$

where

$$F_{effective} = F - R_a - R_r - R_g$$

If the mass factor is ignored (i.e., $\gamma_m = 1$), then:

$$F = ma + R_a + R_r + R_g$$

$$F - R_a - R_r - R_g = ma$$

$$F_{effective} = ma$$

Example 6.4

For a given roadway condition, a truck can apply a maximum of 290 lb force on a flat road. The air resistance of the truck is 2 lb, the rolling resistance is 30 lb. The weight of the truck is 3,000 lb. Consider the mass factor for the inertia of the vehicle's rotating parts. The car's engine is producing an overall gear reduction ratio of 4.5 to 1. Calculate the acceleration of the truck.

SOLUTION

Given,

Maximum force, $F = 290$ lb
Air resistance, $R_a = 2$ lb
Rolling resistance, $R_r = 30$ lb
Grade resistance, $R_g = 0$
Weight of the truck, $W = 3,000$ lb
Overall gear reduction ratio, $\varepsilon_o = 4.5$
Acceleration of the truck, $a = ?$

Known, $F - R_a - R_r - R_g = \gamma_m m a$

Therefore, $a = \dfrac{F - R_a - R_r - R_g}{\gamma_m m}$

The mass factor, $\gamma_m = 1.04 + 0.0025\varepsilon_o^2 = 1.04 + 0.0025(4.5)^2 = 1.09$

$$a = \frac{290\,\text{lb} - 2\,\text{lb} - 30\,\text{lb} - 0\,\text{lb}}{(1.09)\left(\dfrac{3,000\ \text{lb}}{32.2\,\dfrac{\text{ft}}{\text{sec}^2}}\right)}$$

$$= 2.54\ \frac{\text{ft}}{\text{sec}^2}$$

Answer: 2.54 ft/sec²

Example 6.5

A straight roadway segment with an upward grade of 5% is aligned to the north, speed limit of 65 mph, and has a wind speed of 33.63 mph toward the south-west. Assume, the drag coefficient is 0.6 and the air density is 0.00237 slug/ft³. Ignore the mass factor. Calculate the minimum power of a 4,000-lb car with a cross-sectional area of 40 ft² to drive at an acceleration of 3.0 ft/sec².

SOLUTION

Air resistance depends on both vehicle and wind speeds. The vehicle speed and component of wind speed in opposition to the vehicle are to be added together to determine the net speed.

Car speed, $v_{car} = 65$ mph (north)
Wind speed, $v_{wind} = 33.63$ mph (south-west)
Net speed, $v_{net} = 65 + 33.63 \cos 45 = 88.78$ mph (north) (Figure 6.4)

Known, $F - R_a - R_r - R_g = ma$ (ignoring the mass factor)
Or, $F_{effective} = ma$

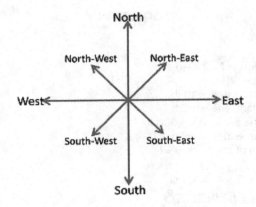

FIGURE 6.4 Directions of Plane

$$F_{effective} = \left(\frac{4,000 \text{ lb}}{32.2 \dfrac{\text{ft}}{\text{sec}^2}} \right) \left(3.0 \dfrac{\text{ft}}{\text{sec}^2} \right) = 373 \text{ lb}$$

Now, let us find out the resistance forces.

$$88.78 \frac{\text{mile}}{\text{hour}} = \frac{88.78 \text{mile} \left(5,280 \dfrac{\text{ft}}{\text{mile}} \right)}{(1 \text{hour}) 3,600 \dfrac{\text{sec}}{\text{hour}}} = 130.2 \frac{\text{ft}}{\text{sec}}$$

$$65 \frac{\text{mile}}{\text{hour}} = \frac{65 \text{mile} \left(5,280 \dfrac{\text{ft}}{\text{mile}} \right)}{(1 \text{hour}) 3,600 \dfrac{\text{sec}}{\text{hour}}} = 95.33 \frac{\text{ft}}{\text{sec}}$$

Weight of the car, $W = 4{,}000$ lb
Cross-sectional area, $A = 40$ ft^2
Acceleration, $a = 3.0$ ft/sec^2
Drag coefficient, $C_D = 0.6$
Air density, $\rho = 0.00237$ slug/ft^3

Air resistance, $R_a = \dfrac{\rho}{2} C_D A_f v^2$

$$= \frac{\left(0.00237 \dfrac{\text{slug}}{\text{ft}^3} \right)}{2} (0.6)\left(40 \text{ ft}^2 \right) \left(130.2 \dfrac{\text{ft}}{\text{sec}} \right)^2$$

$$= 482.2 \text{ lb}$$

Rolling resistance, $R_r = \mu_r(W)$

$$= 0.01\left(1 + \frac{v}{147}\right)(W)$$

$$= 0.01\left(1 + \frac{95.33\,\dfrac{ft}{sec}}{147}\right)(4{,}000\ lb)$$

$$= 65.9\ lb$$

Grade resistance, $R_g = WG$

$$= (4{,}000\ lb)\left(\frac{5}{100}\right)$$

$$= 200\ lb$$

Now, $F_{effective} = F - R_a - R_r - R_g$

$$F = F_{effective} + R_a + R_r + R_g$$

$$= 373\ lb + 482.2\ lb + 65.9\ lb + 200\ lb$$

$$= 1{,}121\ lb$$

Power, $P = Fv$

$$= 1{,}121\ lb\left(95.33\,\frac{ft}{sec}\right)$$

$$= 106{,}878\,\frac{lb.ft}{sec}$$

$$= \frac{106{,}878\,\dfrac{lb.ft}{sec}}{550\,\dfrac{\frac{lb.ft}{sec}}{hp}}$$

$$= 194\ hp$$

Answer: 194 hp
[Recall, 1 hp = 550 lb.ft/sec and 1 hp = 746 W]

6.4 MAXIMUM TRACTIVE EFFORT

The tractive effort available to overcome resistance and/or to accelerate the vehicle is determined either by the force generated by the vehicle's engine or by some maximum value that will be a function of the vehicle's weight distribution and the characteristics of the roadway surface–tire interface. The basic concepts underlying these two determinants of available tractive effort are presented here. No matter how much force a vehicle's engine makes available at the roadway surface, there is a point beyond which additional force merely results in the spinning of tires and does not overcome resistance or accelerate the vehicle. To explain what determines this point of maximum tractive effort (the limiting value beyond which tire spinning begins), a force- and moment-generating diagram is provided in Figure 6.5.

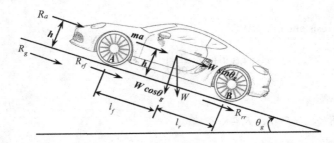

FIGURE 6.5 Vehicle Forces and Moment-Generating Distances R_a=aerodynamic resistance, lb or NR_{rr}=rolling resistance of the rear tires, lb or NR_{rf}=rolling resistance of the front tires, lb or NW_f=weight of the vehicle on the front axle, lb or NW_r=weight of the vehicle on the rear axle, lb or NW=total vehicle weight, lb or NF_r=available tractive effort of the rear tires, lb or NF_f=available tractive effort of the front tires, lb or Nm=vehicle mass, slug or kg$a$=acceleration, ft/sec² or m/sec²θ_g=angle of the grade, degrees$h$=height of the center of gravity above the roadway surface, ft or ml_r=distance from the rear axle to the center of gravity, ft or ml_f=distance from the front axle to the center of gravity, ft or mL=length of wheelbase (l_f+l_r), ft or m

To determine the maximum tractive effort that the roadway surface–tire contact can support, it is necessary to examine the normal loads on the axles. The normal load on the rear axle (W_r) is given by summing the moments about point A (see Figure 6.5):

$$R_a h + \left(W \cos\theta_g \right) l_f + mah \pm \left(W \sin\theta_g \right) h - W_r L = 0$$

$$W_r = \frac{R_a h + Wl_f \cos\theta_g + mah \pm Wh\sin\theta_g}{L}$$

In this equation the grade moment ($Wh \sin\theta_g$) is positive for an upward slope and negative for a downward slope. Rearranging terms (assuming $\cos\theta_g = 1$ for the small grades encountered in highway roadway applications) and substituting into the equation gives

$$W_r = \frac{R_a h + Wl_f + mah \pm h\left(W \sin\theta_g \right)}{L}$$

$$W_r = \frac{Wl_f}{L} + \frac{h}{L}\left(R_a + ma \pm R_g\right)$$

$$W_r = \frac{Wl_f}{L} + \frac{h}{L}\left(R_a + ma \pm R_g\right)$$

Now, $F - R_a - R_r \pm R_g = ma$

Therefore, $W_r = \frac{Wl_f}{L} + \frac{h}{L}\left(R_a + \left(F - R_a - R_r \pm R_g\right) \pm R_g\right)$

$$W_r = \frac{Wl_f}{L} + \frac{h}{L}\left(R_a + \left(F - R_a - R_r \pm R_g\right) \pm R_g\right)$$

$$W_r = \frac{l_f}{L}W + \frac{h}{L}\left(F - R_r\right)$$

R_g cancels each other regardless of sign. The maximum tractive effort, as determined by the roadway surface–tire interaction, will be the normal force multiplied by the coefficient of road adhesion (μ). Generally, the road adhesion coefficient is the maximum value of road friction coefficient (μ_r), and in some researches, friction coefficient is used to represent the maximum value. It is the ratio of the tractive effort force just necessary to slip the wheels on the track to the adhesive weight. It reduces with increase in the speed.

For a rear-wheel-drive car: $F_{max,rear} = \mu W_r$

$$F_{max,rear} = \mu\left[\frac{l_f}{L}W + \frac{h}{L}\left(F - R_r\right)\right]$$

$$F_{max,rear} = \frac{\dfrac{\mu W}{L}\left(l_f - \mu_r h\right)}{1 - \dfrac{\mu h}{L}}$$

where,
   μ_r = coefficient of rolling resistance
   μ = coefficient of road adhesion

Similarly, by summing moments about point B, it can be shown that for a front-wheel-drive vehicle:

$$F_{max,front} = \frac{\dfrac{\mu W}{L}\left(l_r - \mu_r h\right)}{1 + \dfrac{\mu h}{L}}$$

Because of canceling of units, h, l_f, l_r, and L can be in any unit of length (ft, in., etc.). However, all of these terms must be in the same chosen unit of measure. The units of F_{max} will be the same as the units for W.

Example 6.6

A 5,000-lb car is designed with a 120-in. wheelbase. The center of gravity is located 22 in. above the pavement and 40 in. behind the front axle. If the coefficient of road adhesion is 0.7, calculate the maximum tractive effort that can be developed if the car is:

    (a) front-wheel-drive, and
    (b) rear-wheel-drive.

SOLUTION

Given,

    Coefficient of road adhesion, $\mu = 0.7$
    Height of center of gravity, $h = 22$ in.
    Length of wheelbase, $L = 120$ in.
    Weight of car, $W = 5,000$ lb
    Distance from the front axle to the center of gravity, $l_f = 40$ in.
    Distance from the rear axle to the center of gravity, $l_r = 120$ in. $-$ 40 in. $= 80$ in.

Given, velocity is zero. Therefore, coefficient of rolling resistance

$$\mu_r = 0.01\left(1 + \frac{v}{147}\right)$$

$$= 0.01\left(1 + \frac{0}{147}\right)$$

$$= 0.01$$

    (a) For the front-wheel-drive:

$$F_{max,front} = \frac{\frac{\mu W}{L}\left(l_r - \mu_r h\right)}{1 + \frac{\mu h}{L}}$$

$$= \frac{\frac{(0.7)(5,000\,\mathrm{lb})}{120\,\mathrm{in.}}\left(80\,\mathrm{in.} - 0.01(22\,\mathrm{in.})\right)}{1 + \frac{(0.7)(22\,\mathrm{in.})}{120\,\mathrm{in.}}}$$

$$= 2,062\ lb$$

(b) For the rear-wheel-drive:

$$F_{max,rear} = \frac{\frac{\mu W}{L}(l_f - \mu_r h)}{1 - \frac{\mu h}{L}}$$

$$= \frac{\frac{(0.7)(5,000\,lb)}{120\,in.}\left(40\,in. - (0.01)(22\,in.)\right)}{1 - \frac{(0.7)(22\,in.)}{120\,in.}}$$

$$= 1,331\,lb$$

Answers:

(a)   *2,062 lb*
(b)   *1,331 lb*

6.5 STOPPING DISTANCES

6.5.1 THEORETICAL DISTANCE

Stopping distance includes two phases of stopping a vehicle: Perception-reaction time and braking time. Perception-reaction time is the time a driver takes to realize that an action is essential for a road condition, decide to apply the brake, and actually apply the brake. Braking time is the time to complete deceleration and stop.

If a brake force (F_b) is applied, F_b will be opposite of the motion and the acceleration a will be negative as F_b will resist the movement and deceleration will occur instead of acceleration. By looking at Figure 6.6, it can be seen that the relationship

$$-F_b = \gamma_b m(-a) + R_a + R_r + R_g$$

$$F_b + R_a + R_r + R_g = \gamma_b ma$$

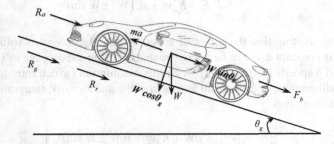

FIGURE 6.6   Forces Acting on a Vehicle during Braking. *Photo courtesy of drawcarz .com*

The mass factor accounting for moments of inertia during braking is denoted by γ_b. The mass factor accounting for moments of inertia during braking, γ_b, has the value of 1.04 for automobiles (Wong 2008).

$$F_b + R_a + R_r + R_g = \gamma_b ma$$

$$a = \frac{F_b + R_a + R_r + R_g}{\gamma_b m}$$

We know that $ads = vdv$

$$ds = \frac{vdv}{a}$$

$$\int ds = \int_{v_1}^{v_2} \frac{vdv}{a}$$

$$S = \int_{v_1}^{v_2} \frac{vdv}{\left(\dfrac{F_b + R_a + R_r + R_g}{\gamma_b m}\right)}$$

$$S = \gamma_b m \int_{v_1}^{v_2} \frac{vdv}{F_b + R_a + R_r + R_g}$$

We know, $R_a = \dfrac{\rho}{2} C_D A_f v^2$. Let us assume $K_a = \dfrac{\rho}{2} C_D A_f$ for simplicity. Then, $R_a = K_a v^2$

Therefore, the equation of S can be written as follows:

$$S = \gamma_b m \int_{v_1}^{v_2} \frac{vdv}{F_b + K_a v^2 + \mu_r (W) \pm W \sin\theta_g}$$

Continuing, assume that the effect of speed on the coefficient of rolling resistance, μ_r, is constant and can be approximated by using the average of initial (v_1) and final (v_2) speeds $[v = (v_1 + v_2)/2]$. With this assumption (which introduces only a very small amount of error), and letting $m = W/g$ and $F_b = \mu W$, integration of the above equation gives

$$S = \frac{\gamma_b W}{2gK_a} \ln\left[\frac{\mu W + K_a v_1^2 + \mu_r W \pm W \sin\theta_g}{\mu W + K_a v_2^2 + \mu_r W \pm W \sin\theta_g}\right]$$

where μ is the coefficient of road adhesion and g is the gravitational constant (32.2 ft/sec^2). If the vehicle is assumed to stop ($v_2=0$) and $v_1=v$, which is the target for the stopping sight distance, then the equation of S can be written as follows:

$$S = \frac{\gamma_b W}{2gK_a} \ln \left[\frac{\mu W + \mu_r W \pm W \sin \theta_g}{\mu W + \mu_r W \pm W \sin \theta_g} + \frac{K_a v^2}{\mu W + \mu_r W \pm W \sin \theta_g} \right]$$

$$S = \frac{\gamma_b W}{2gK_a} \ln \left[1 + \frac{K_a v^2}{\mu W + \mu_r W \pm W \sin \theta_g} \right]$$

True optimal brake force proportioning is seldom achieved in standard non-antilock braking systems, it is useful to define a braking-efficiency term that reflects the degree to which the braking system is operating below optimal. Simply stated, braking efficiency is defined as the ratio of the maximum rate of deceleration, expressed in g's (g_{max}), achievable prior to any wheel lockup to the coefficient of road adhesion:

$$\eta_b = \frac{g_{max}}{\mu}$$

where

η_b = braking efficiency
g_{max} = maximum deceleration in g units (with the absolute maximum = μ)
μ = coefficient of road adhesion

With braking efficiency considered, the actual braking force is, $F_b = \eta_b \mu W$. Therefore, by substitution, the theoretical stopping distance (S) can be analytically calculated as follows:

$$S = \frac{\gamma_b W}{2gK_a} \ln \left[1 + \frac{K_a v^2}{\eta_b \mu W + \mu_r W \pm W \sin \theta_g} \right]$$

where

γ_b = mass factor accounting for moments of inertia during braking, 1.04 for automobiles
v_1 = initial vehicle speed (fps, m/sec)
$K_a = \frac{\rho}{2} C_D A_f$
μ = coefficient of road adhesion
μ_r = frictional coefficient
η_b = braking efficiency
$W \sin \theta_g$ is positive for upward grade and negative for downward grade

If aerodynamic resistance is ignored (due to its comparatively small contribution to braking), integration gives the theoretical stopping distance as:

$$S = \frac{\gamma_b \left(v_1^2 - v_2^2\right)}{2g\left(\eta_b \mu + \mu_r \pm \sin\theta_g\right)}$$

$$S = \frac{\gamma_b v^2}{2g\left(\eta_b \mu + \mu_r \pm \sin\theta_g\right)} \quad [\text{if } v_2 = 0 \text{ and express } v_1 = v]$$

Example 6.7

A new experimental 5,000-lb car, with $C_D = 0.25$ and $A_f = 18$ ft², is traveling at 90 mph up a 10% grade. The coefficient of road adhesion is 0.7 and the air density is 0.0024 slugs/ft³. The car has an advanced antilock braking system that gives it a braking efficiency of 100%. Determine the theoretical minimum stopping distance for the case where aerodynamic resistance:

(a) is considered
(b) is ignored

SOLUTION

Mass factor accounting for moments of inertia during braking, $\gamma_b = 1.04$

The grade is +10%. Therefore, $\theta_g = \tan^{-1}\left(\frac{10}{100}\right) = 5.71°$

Average speed for the coefficient of rolling resistance, $v = (90 \text{ mph} + 0 \text{ mph})/2 = 45$ mph

Coefficient of rolling resistance, $\mu_r = 0.01\left(1 + \frac{v}{147}\right)$

$$= 0.01\left(1 + \frac{45\,\text{mph}\left(1.47\,\dfrac{\text{fps}}{\text{mph}}\right)}{147}\right)$$

$$= 0.0145$$

[1 mph = 1.47 fps]

(a) For the case where aerodynamic resistance is considered:

$$K_a = \frac{\rho}{2}C_D A_f$$

$$= \frac{0.0024}{2}(0.25)(18)$$

$$= 0.0054$$

$$S = \frac{\gamma_b W}{2gK_a} \ln\left[1 + \frac{K_a v^2}{\eta_b \mu W + \mu_r W \pm W \sin\theta_g}\right]$$

$$S = \frac{(1.04)(5,000\,\text{lb})}{2\left(32.2\dfrac{\text{ft}}{\text{sec}^2}\right)(0.0054)}$$

$$\times \ln\left[1 + \frac{(0.0054)\left(90 \times 1.47\dfrac{\text{ft}}{\text{sec}}\right)^2}{(1.0)(0.7)(5,000\,\text{lb}) + (0.0145)(5,000\,\text{lb}) + (5,000\,\text{lb})\sin 5.71}\right]$$

$$S = 14,952.84\ln\left[1 + \frac{94.52}{3,500 + 72.5 + 497.47}\right]$$

Therefore, $S = 343$ ft

(b) For the case where aerodynamic resistance is ignored:

$$S = \frac{\gamma_b v^2}{2g\left(\eta_b \mu + \mu_r \pm \sin\theta_g\right)}$$

$$S = \frac{(1.04)\left(90 \times 1.47\dfrac{\text{ft}}{\text{sec}}\right)^2}{2\left(32.2\dfrac{\text{ft}}{\text{sec}^2}\right)\left((1.0)0.7 + 0.0145 + \sin 5.71\right)}$$

$$S = 347\,\text{ft}$$

Answers:

(a) *343 ft*
(b) *347 ft*

6.5.2 PRACTICAL STOPPING DISTANCE

The stopping distance (S) is the distance required to safely stop a vehicle to avoid a collision with an unexpected stationary object on the roadway ahead for a given design speed. The theoretical assessment of vehicle stopping distance presented in the previous section provided the principles of braking for an individual vehicle under specified roadway surface conditions. However, highway engineers face a more complex problem because they must design for a variety of driver skill levels (which can affect whether or not the brakes lock and reduce the coefficient of road adhesion to slide values), vehicle types (with varying aerodynamics, weight distributions, and brake efficiencies), and weather conditions (which change the roadway's coefficient of adhesion). As a result of the wide variability inherent in the determination of braking distance, an equation is required that provides an estimate of typical observed braking distances and is more simplistic. The basic physics equation on rectilinear motion, assuming constant deceleration, is chosen as the basis of a practical equation for stopping distance:

$$v_2^2 = v_1^2 + 2ad$$

where
v_2 = final vehicle speed, ft/sec
v_1 = initial vehicle speed in ft/sec
a = acceleration (negative for deceleration), ft/sec^2
d = deceleration distance (practical stopping distance), ft

Rearranging the equation and assuming a is negative for deceleration gives

$$d = \frac{v_1^2 - v_2^2}{2a}$$

If $v_2 = 0$ (the vehicle comes to a complete stop), the practical stopping distance equation is:

$$d = \frac{v_1^2}{2a}$$

To make this equation generally applicable for design purposes, a deceleration rate, a, must be chosen that is representative of appropriately conservative braking behavior. AASHTO (2018) recommends a deceleration rate of 11.2 ft/sec^2. Empirical studies (Fambro et al. 1997) have shown that approximately 90% of drivers decelerate at rates greater than this, and that this deceleration rate is well within a driver's capability to maintain steering control during a braking maneuver on wet surfaces. Additionally, empirical studies (Fambro et al. 1997) have confirmed that most vehicle braking systems and tire-pavement friction levels are capable of supporting this deceleration rate, even under wet conditions. To account for the effect of grade, the above can be modified as follows:

$$d = \frac{v_1^2}{2g\left(\dfrac{a}{g} \pm G\right)}$$

where
g = gravitational constant, 32.2 ft/sec^2
G = roadway grade (+ for uphill, − for downhill) in percent/100
Other terms are as defined previously

If we use v_i as v and the unit of v as miles per hour, then:

$$d = \frac{v^2}{2g\left(\dfrac{a}{g} \pm G\right)} \quad [v \text{ is in fps}]$$

$$d = \frac{(1.47v)^2}{2(32.2)\left(\dfrac{a}{32.2} \pm G\right)} \quad [v \text{ is in mph. } 1 \text{ mph} = 1.47 \text{ fps}]$$

$$d = \frac{v^2}{30\left(\dfrac{a}{32.2} \pm G\right)} \quad [v \text{ is in mph.}]$$

Until now the focus has been directed toward the distance required to stop the vehicle from the point of brake application. However, in providing sufficient sight distance for a driver to stop safely, it is also necessary to consider the distance traveled during the time the driver is perceiving and reacting to the need to stop. The distance traveled during perception/reaction (d_r) is given by

$$d_r = vt \quad [v \text{ is in ft/sec}]$$

$$d_r = 1.47vt \quad [v \text{ is in mph.}]$$

where
v = initial vehicle speed
t = time required to perceive and react to the need to stop, sec

The perception/reaction time of a driver is a function of a number of factors, including the driver's age, physical condition, and emotional state, as well as the complexity of the situation and the strength of the stimuli requiring a stopping action. For highway design, a conservative perception/reaction time has been determined to be 2.5 seconds (AASHTO 2018). For comparison, average drivers have perception/reaction times of approximately 1.0 to 1.5 seconds. Thus, the total required stopping distance is a combination of the braking distance and the distance traveled during perception/reaction as:

$$S = d_r + d$$

S = total stopping distance (including perception/reaction), ft
d = distance traveled during braking, ft
d_r = distance traveled during perception/reaction, ft

Therefore, S can reasonably be calculated using the following equation:

$$S = 1.47vt + \frac{v^2}{30\left(\dfrac{a}{32.2} \pm G\right)}$$

where
S= stopping sight distance, ft [no other unit is possible]
t= driver perception-reaction time (sec), commonly used as 2.5 sec
v= vehicle approach speed, mph [no other unit is possible]
G= percent grade divided by 100 (uphill grade '+', downhill grade '–')
a= deceleration rate (ft/sec^2) [11.2 ft/sec^2 may be used if not reported]

The American Association of State Highway and Transportation Officials (AASHTO) recommends 1.5 seconds for perception time and 1.0 second for reaction time, although this time varies with driver's age, time of day, driving history, etc.

Example 6.8

The design speed of a highway is 103 mph with an uphill grade of 1%. The driver reaction time is 2.5 sec for that highway. If a car sees an object ahead and decelerates at 30 ft/sec^2, calculate the stopping sight distance for that car.

SOLUTION

Given,

Speed, v = 103 mph
Grade, G = 1% = 0.01
Driver reaction time, t = 2.5 sec
Deceleration, a = 30 ft/sec^2
Stopping sight distance, S = ?

$$\text{Stopping sight distance, } S = 1.47vt + \frac{v^2}{30\left(\dfrac{a}{32.2} + G\right)}$$

$$= 1.47(103\,\text{mph})(2.5\,\text{sec}) + \frac{(103\,\text{mph})^2}{30\left(\dfrac{30\dfrac{\text{ft}}{\text{sec}^2}}{32.2} + 0.01\right)}$$

$$= 754\ \text{ft}$$

Answer: 754 ft
Note: Be careful about the unit. v is in mph. However, S is in ft.

Example 6.9

Two drivers each have a perception-reaction time of 2.5 sec. One is obeying a 55-mph speed limit and the other is traveling illegally at 70 mph. The road has a downgrade of 2.5%. Calculate the total stopping distance for each driver using the practical stopping distance equation.

SOLUTION

Given,

Grade, $G = -2.5\% = -0.025$
Driver reaction time, $t = 2.5$ sec
Deceleration, $a = 11.2$ ft/sec^2 [default value]
Stopping sight distance, $S = ?$

For the driver traveling at speed, $v = 55$ mph:

$$S = 1.47vt + \frac{v^2}{30\left(\dfrac{a}{32.2} \pm G\right)}$$

$$= 1.47(55\text{mph})(2.5\text{sec}) + \frac{(55\text{mph})^2}{30\left(\dfrac{11.2}{32.2} - 0.025\right)}$$

$$= 201.13 + \frac{6,507}{20.79}\text{ft}$$

$$= 514.1\text{ft}$$

For the driver traveling at speed, $v = 70$ mph:

$$S = 1.47vt + \frac{v^2}{30\left(\dfrac{a}{32.2} \pm G\right)}$$

$$= 1.47(70\text{mph})(2.5\text{sec}) + \frac{(70\text{mph})^2}{30\left(\dfrac{11.2}{32.2} - 0.025\right)}$$

$$= 257.25 + 505.95\text{ft}$$

$$= 763.2\text{ft}$$

Therefore, driving at 70 mph increases the total stopping distance by a very substantial 763.2 ft – 514.1 ft = 249.1 ft.
Answers:

514 ft, 763 ft

6.6   VEHICLE DYNAMICS IN HORIZONTAL CURVES

When a vehicle runs around a circular curve (Figure 6.7), two forces are generated: inward centripetal and outward centrifugal forces. Centripetal force acts radially inward along the radius of curvature.

FIGURE 6.7   A Circular Curve

Let us assume a particle is moving along a circular path at a constant speed, as shown in Figure 6.8, Step 1. At any instant, its velocity is $v(1)$. After some period, it moves to another point with velocity $v(2)$. Both the velocities are numerically equal with different directions. Newton's First Law says without applying forces, the motion of a body does not change. Therefore, there must be a force that exists in this circular motion which is causing this directional change of velocity. The acceleration due to the change in velocity direction acts along the center and is known as normal component of acceleration. The normal component of acceleration which always acts toward to center of curvature can be calculated as:

$$a_n = \frac{v^2}{R}$$

where

$v =$ velocity of the object in ft/sec. This velocity works along the tangent of the curve. If the magnitude of the velocity does not change over time, then there is no tangential component of the acceleration. However, there is always the normal component of acceleration in a curved or circular path due to the change in direction of the velocity.

$R =$ radius of curvature in ft

Then, the centripetal force can be calculated as:

$$F_{centripetal} = \frac{mv^2}{R}$$

where $m=$ mass of the object in slug. Now, let us see Step 2 of the vector analysis. The velocities are transferred to make a triangle. Then, following the arrows of the triangle, $v(1)+ v(3)=v(2)$ or $v(3)=v(1) - v(2)$. This means another velocity, $v(3)$, is acted on the particle which is causing this directional change. The force which is creating the third velocity, $v(3)$, must act along the direction of the third velocity, $v(3)$. Step 3 shows that the direction of the third velocity, $v(3)$ is along the center of the circle. That force, which is creating the third velocity $v(3)$, is along the center, and is called centripetal (center-loving) force.

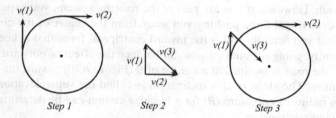

Step 1   Step 2   Step 3

FIGURE 6.8 Change of Velocity in a Circular Path

Now, let us talk about how this centrifugal force (away from center) arises. To change the direction of velocity, centripetal force acts along the center. The part of the body (say, leg of a person, tire of a car, etc.) which is in contact with the path is attracted toward the center due to this centripetal force. However, the upper part of the body tends to stay in position due to the inertia, as shown in Figure 6.9, apparently gets centrifugal force and tends to be thrown away from center. Therefore, centripetal force and centrifugal force are numerically equal but opposite in direction as:

$$F_{centripetal} = F_{centrifugal} = \frac{mv^2}{R}$$

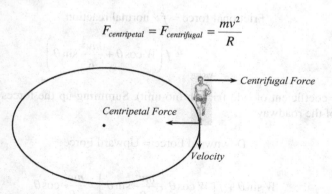

Centrifugal Force

Centripetal Force

Velocity

FIGURE 6.9 Development of Centrifugal Force in a Circular Path

Some people may confuse centrifugal force with its counterpart, centripetal force. This is because they are closely related like the two sides of a coin. Centripetal force is the component of force acting on a body in a curvilinear motion directed toward the center of curvature. While centrifugal force is the apparent force, equal and opposite to the centripetal force, caused by the inertia of the body. Centripetal force draws a rotating body away from the center of rotation.

Note that while centripetal force is an actual force, centrifugal force is defined as an apparent force. In other words, when twirling a mass on a string, the string exerts an inward centripetal force on the mass, while mass appears to exert an outward force on the string. If you are observing a rotating system from the outside, you see an inward centripetal force acting to constrain the rotating body to a circular path. However, if you are part of the rotating system, you experience an apparent centrifugal force pushing you away from the center of the circle, even though what you actually feel is the inward centripetal force that is keeping you from, literally, going off on a tangent. To balance the effect of outward centrifugal force, the road is inclined at an angle of θ (Figure 6.10) toward the center of the curvature. The slope of this inclination is called the superelevation (e). The minimum radius of curvature (R) for a circular motion can be determined using the following relationship:

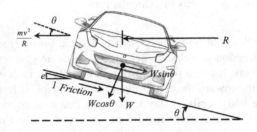

FIGURE 6.10   A Vehicle in a Curve with Lateral Inclination

$$\text{Frictional force} = f \times \text{normal reaction}$$

$$= f\left(W\cos\theta + \frac{mv^2}{R}\sin\theta \right)$$

where f=coefficient of side friction (no unit). Summing up the forces along the surface of the roadway:

$$\text{Downward Force} = \text{Upward Force}$$

$$W\sin\theta + f\left(W\cos\theta + \frac{mv^2}{R}\sin\theta \right) = \frac{mv^2}{R}\cos\theta$$

$$W\sin\theta + f\left(W\cos\theta + \frac{W}{g}\frac{v^2}{R}\sin\theta \right) = \frac{W}{g}\frac{v^2}{R}\cos\theta$$

where
   v = vehicle speed, ft/sec
   g = gravitational constant, 32.2 ft/sec^2
   W = weight of the vehicle, lb
   θ = angle of incline, degrees

Dividing both sides of the equation by $W\cos\theta$ gives

$$\frac{W\sin\theta}{W\cos\theta} + f\left(\frac{W\cos\theta}{W\cos\theta} + \frac{\dfrac{W}{g}\dfrac{v^2}{R}\sin\theta}{W\cos\theta}\right) = \frac{\dfrac{W}{g}\dfrac{v^2}{R}\cos\theta}{W\cos\theta}$$

$$\tan\theta + f\left(1 + \frac{v^2}{gR}\tan\theta\right) = \frac{v^2}{gR}$$

$$\tan\theta + f + f\frac{v^2}{gR}\tan\theta = \frac{v^2}{gR}$$

$$\tan\theta + f = \frac{v^2}{gR} - f\frac{v^2}{gR}\tan\theta$$

$$\tan\theta + f = \frac{v^2}{gR}\left(1 - f\tan\theta\right)$$

The term $\tan\theta$ indicates the superelevation of the curve (banking) and can be expressed in percent; it is denoted as e ($e = 100\tan\theta$). In words, the superelevation is the number of vertical feet (meters) of rise per 100 feet (meters) of horizontal distance. The term $f\tan\theta$ in equation form is conservatively set equal to zero for practical applications due to the small values that f and θ typically assume (this is equivalent to ignoring the normal component of centripetal force). With $e = 100\tan\theta$, the equation can be arranged as:

$$e + f = \frac{v^2}{gR}\left(1 - f\tan\theta\right)$$

$$e + f = \frac{v^2}{gR}\left(1 - 0\right)$$

$$e + f = \frac{v^2}{gR}$$

Now, if v is used in mph [1 mph = 1.47 fps], then:

$$e + f = \frac{(1.47v)^2}{\left(32.2\,\frac{\text{ft}}{\text{sec}^2}\right)R}$$

$$e + f = \frac{v^2}{15R}$$

where
   $e = \tan\theta =$ superelevation (in decimal, 1% →0.01)
   $f =$ coefficient of side friction
   $v =$ vehicle speed, mph [no other unit is possible]
   $R =$ radius of curvature, ft [no other unit is possible]

The above equation can be derived by equating the component of centrifugal force along the curved roadway, and the resisting friction force and the weight component along the curved roadway. Figure 6.11 dictates the imp ortance of adequate curve length where a large vehicle cannot maneuver around a sharp curve.

FIGURE 6.11   Inadequate Sharp Curve (*Courtesy of azfamily*)

Example 6.10

A highway is to be designed for a speed of 88 mph and superelevation of 6%. If the coefficient of side friction is 0.2, calculate the radius of curvature of the horizontal curve required.

SOLUTION

Given,
   Speed, $v = 88$ mph
   Superelevation, $e = 6\%$

Coefficient of side friction, $f=0.2$

Radius of curvature of the horizontal curve, $R = ?$

Known, $e+f = \dfrac{v^2}{15R}$

$$0.06+0.2 = \dfrac{(88\,\text{mph})^2}{15R}$$

Therefore, $R=1{,}986$ ft $\approx 2{,}000$ ft

Answer: 2,000 ft

Note: Be careful about the unit. v is in mph; e and f are in decimals, and R is in ft.

The side friction factor (f) at impending skid depends on a number of other factors, among which the most important are the speed of the vehicle, the type and condition of the roadway surface, and the type and condition of the vehicle tires. The maximum side friction factors developed between new tires and wet concrete pavements range from about 0.5 at 30 kmph (20 mph) to approximately 0.35 at 100 kmph (60 mph). For normal wet concrete pavements and smooth tires, the maximum side friction factor at impending skid is about 0.35 at 70 kmph (45 mph). In all cases, the studies show a decrease in side friction values as speeds increase. One series of tests found coefficients of friction for ice ranging from 0.05 to 0.20, depending on the condition of the ice (i.e., wet, dry, clean, smooth, or rough). Tests on loose or packed snow show coefficients of friction ranging from 0.20 to 0.40 (AASHTO 2018).

The maximum rates of superelevation used on highways are controlled by four factors:

(1) Climate conditions (i.e., frequency and amount of snow and ice)
(2) Terrain conditions (i.e., flat, rolling, or mountainous)
(3) Type of area (i.e., rural or urban)
(4) Frequency of very slow-moving vehicles

The highest superelevation rate for highways in common use is 10%, although 12% is used in some cases. Superelevation rates above 8% are only used in areas without snow and ice. Although higher superelevation rates offer an advantage to those drivers traveling at high speeds, current practice considers that rates in excess of 12% are beyond practical limits (AASHTO 2018).

The minimum radius is a limiting value of curvature for a given design speed and is determined from the maximum rate of superelevation and the maximum side friction factor selected for design. The minimum radius of curvature is based on a threshold of driver comfort that is sufficient to provide a margin of safety against skidding and vehicle rollover.

FUNDAMENTALS OF ENGINEERING (FE) EXAM STYLE QUESTIONS

FE Problem 6.1

A ball attached to a string is whirled around in a horizontal circle having a radius of R. If the radius of the circle is changed to $4R$ and the same centripetal force is applied by the string, the new speed of the ball is which of the following?

    A. one-quarter the original speed
    B. one-half the original speed
    C. the same as the original speed
    D. twice the original speed

FE Problem 6.2

A racing car is moving around the circular track of radius 300 m shown in Figure 6.12. At the instant when the car's velocity is constant and directed due east, its acceleration is directed due south and has a magnitude of 3 m/sec². When viewed from above, the car is moving:

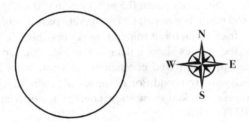

FIGURE 6.12

    A. clockwise at 30 m/sec
    B. clockwise at 10 m/sec
    C. counterclockwise at 30 m/sec
    D. counterclockwise at 10 m/sec

FE Problem 6.3

An automobile moves at a constant speed down one hill and up another hill along a smoothly curved surface as shown in Figure 6.13. Which of the following diagrams best represents the directions of the velocity (v) and the acceleration (a) of the automobile at the instant that it is at the lowest position as shown?

FIGURE 6.13

FE Problem 6.4

A car initially travels north and then turns to the left along a circular curve. This causes a package on the seat of the car to slide toward the right side of the car. Which of the following is true of the net force on the package while it is sliding?

A. The force is directed away from the center of the circle.
B. There is not enough force directed north to keep the package from sliding.
C. There is not enough force tangential to the car's path to keep the package from sliding.
D. There is not enough force directed toward the center of the circle to keep the package from sliding.

FE Problem 6.5

A child accidentally runs into the street in front of an approaching vehicle. The vehicle is traveling at 45 mph. Assuming the road is level, at what distance in ft must the driver first see the child to stop just in time?

A. 165
B. 195
C. 360
D. 425

FE Problem 6.6

Aerodynamic resistance is most nearly:

A. directly proportional to vehicle speed
B. directly proportional to air speed
C. directly proportional to the square of the net vehicle speed with respect to air
D. inversely proportional to the square of the net vehicle speed with respect to air

FE Problem 6.7

Grade resistance R_g occurs due to:

A. transverse slope
B. longitudinal slope

C. lateral slope
D. roadside shoulder slope

FE Problem 6.8
Surface Resistance/Rolling Resistance in cars occur primarily due to:

A. surface roughness
B. speed
C. longitudinal slope
D. transverse slope

FE Problem 6.9
The maximum rates of superelevation used on highways are controlled by:
[select all that apply]

A. climate conditions (i.e., frequency and amount of snow and ice)
B. terrain conditions (i.e., flat, rolling, or mountainous)
C. type of area (i.e., rural or urban)
D. frequency of very slow-moving vehicles

FE Problem 6.10
Centrifugal force on roadways is commonly resisted by:

A. friction between roadways and tires
B. raising the outside of the road
C. both of the options
D. any one of the options

FE Problem 6.11
The 'S' and 'v' in the stopping sight distance equation of $S = 1.47vt + \dfrac{v^2}{30\left(\dfrac{a}{32.2}+G\right)}$

have the units of:

A. mile, miles per hour, respectively
B. feet, miles per hour, respectively
C. mile, feet per hour, respectively
D. feet, feet per hour, respectively

FE Problem 6.12
Upon applying a full brake, a car commonly decelerates at the rate of (ft/sec²):

A. 11.2
B. 3.4

C. 30
D. 32.2

FE Problem 6.13

The American Association of State Highway and Transportation Officials (AASHTO) recommended perception time (sec) is most nearly:

A. 1.5
B. 1.0
C. 2.5
D. none of these options

FE Problem 6.14

The side friction factor (μ, sometimes used as f) at impending skid depends on: [select all that apply]

A. the speed of the vehicle
B. the type of the roadway surface
C. the condition of the roadway surface
D. the type and condition of the vehicle tires

FE Problem 6.15

Current practice considers that superelevation rates in horizontal curves in excess of .... are beyond practical limits.

A. 12%
B. 10%
C. 8%
D. 6%

FE Problem 6.16

The weight of a truck is 8,000 lb. It is running at a speed of 60 mph in a 4% upward grade roadway. The surface/rolling resistances (lb) to this truck is most nearly (rounded to nearest 10):

A. 90
B. 110
C. 130
D. 150

FE Problem 6.17

The weight of a truck is 8,000 lb. It is running at a speed of 60 mph in a 4% upward grade roadway. The grade resistances (lb) to this truck is most nearly:

A. 230
B. 320
C. 540
D. 480

FE Problem 6.18

A highway is to be designed with a speed of 50 mph and the superelevation of 4%. If the coefficient of side friction is 0.25, the radius of curvature (ft) of the horizontal curve required is most nearly:

A. 375
B. 475
C. 575
D. 675

FE Problem 6.19

A stone is whirled in a vertical circle, the tension in the string is at maximum:

A. when the string is horizontal
B. when the stone is at the highest position
C. when the stone is at the lowest position
D. at all the positions

Practice Problems

Problem 6.1

A train is traveling toward the north at 50 mph. At that moment, wind is blowing 60 mph toward the south. The projected area of the face of the train is about 75 ft². Assume, the drag coefficient is 1.1, and the air density is 0.00237 slug/ft³. What is most nearly the force exerted by the wind on the train?

Problem 6.2

A train is traveling toward the north at 50 mph. At that moment, wind is blowing 60 mph toward the south-west. The projected area of the face of the train is about 75 ft². Assume, the drag coefficient is 1.1, and the air density is 0.00237 slug/ft³. What is most nearly the force exerted by the wind on the train?

Problem 6.3

The weight of a truck is 8,000 lb. If it runs at a speed of 60 mph on a roadway as listed below, what are the surface/rolling resistances to this truck?

(a) 4% upward grade
(b) 3% downward grade
(c) Flat or near flat

Problem 6.4

The weight of a truck is 8,000 lb. If it runs at a speed of 60 mph on a roadway as listed below, what are the grade resistances to this truck?

(a) 4% upward grade
(b) 3% downward grade
(c) Flat or near flat

Problem 6.5

A straight roadway segment with a downgrade of 5% is aligned to the north, speed limit of 65 mph, and has a wind speed of 30 mph toward the north-west. What should be the minimum horsepower of an 11,000-lb car with the cross-sectional area of 80 ft² to drive at an acceleration of 5.0 ft/sec²? Ignore the mass factor accounts for the inertia of the vehicle's rotating parts. Assume, the drag coefficient is 0.7, and the air density is 0.00237 slug/ft³.

Problem 6.6

The design speed of a highway is 103 mph with an uphill grade of 1%. The driver reaction time is 2.5 sec for that highway. If a car sees an object ahead and decelerates at 30 ft/sec², what will be most nearly the stopping sight distance for that car?

Problem 6.7

A highway is to be designed with a speed of 50 mph and a superelevation of 4%. If the coefficient of side friction is 0.25, what is most nearly the radius of curvature of the horizontal curve required?

Problem 6.8

A straight roadway segment with an upgrade of 5% is aligned to the north, speed limit of 65 mph, and has a wind speed of 30 mph toward the south-west. What should be the minimum horsepower of an 11,000-lb car with the cross-sectional area of 80 ft² to drive at a constant speed? Assume, the drag coefficient is 0.7, and the air density is 0.00237 slug/ft³.

Problem 6.9

A car is traveling at 80 mph and has a braking efficiency of 80%. The brakes are applied to miss an object that is 150 ft from the point of brake application, and the coefficient of road adhesion is 0.85. Ignoring aerodynamic resistance and assuming the theoretical minimum stopping distance, estimate how fast the car will be going when it strikes the object if:

(a) the surface is level, and
(b) the surface is on a 5% upgrade.

REFERENCES

AASHTO 2018. A Policy on Geometric Design of Highways and Streets, 7th ed, 2018, by the American Association of State Highway and Transportation Officials (AASHTO), Washington, DC.

Fambro, D. B., Fitzpatrick, K., and Koppa, R. J. 1997. *Determination of Stopping Sight Distances*. NCHRP Report 400. Transportation Research Board, Washington, DC.

Mannering, F. and Washburn, S. 2013. *Principles of Highway Engineering and Traffic Analysis*, 5th edition. John Wiley & Sons, Inc., Hoboken, NJ.

Wong, J. Y. 2008. *Theory of Ground Vehicles*, 4th edition. John Wiley & Sons, New York.

7 Vibration

7.1 GENERAL

Mechanical vibration, or simply vibration, is the measurement of periodic oscillations with respect to an equilibrium condition of an object. Sometimes, vibrations are desirable in certain types of machine tools or production lines. However, most of the time, vibration of mechanical systems is not desirable due to wasting energy, reducing efficiency, and safety considerations. For example, passenger ride comfort in aircraft or automobiles is greatly affected by the vibrations caused by outside disturbances, such as aero-elastic effects or rough road conditions. In other cases, eliminating vibrations may save human lives; a good example of this is the vibration of civil engineering structures during an earthquake.

Vibrations can be classified as:

(a) *Free and Forced Vibration.* Free vibration means the system is left to vibrate on its own after the initial disturbance. In forced vibration, the system is subjected to an external repeating type of force, and the system vibrates in response to that.

(b) *Damped and Undamped.* If no energy is lost or dissipated by friction or other resistance during oscillation, the vibration is known as undamped vibration. If any energy is lost in this way, it is called damped vibration. In many physical systems, the amount of damping is so small that it can be disregarded for most engineering purposes.

(c) *Linear and Nonlinear.* If all the basic components of a vibration system such as spring, mass, and the damper behave linearly, then it is called linear vibration. In nonlinear vibration, one or more basic components of a vibration system do not behave in a linear manner.

(d) *Deterministic or Not.* Deterministic means the magnitude of the excitation (force or motion) acting on a vibratory system is known at any given time. In random vibration, the value of the excitation at any given time cannot be predicted. Examples of random vibration are wind velocity, road roughness, and ground motion during an earthquake, etc.

(e) *Torsional Vibrations.* Torsional vibrations are angular vibrations of an object, typically a shaft along its axis of rotation. Torsional vibrations are evaluated as the variation of rotational speed within a rotation cycle.

The basic vibrations such as the undamped-free vibration, torsional free vibration, and damped vibration are the subject of this chapter.

DOI: 10.1201/9781003283959-7

7.2   UNDAMPED-FREE LINEAR VIBRATION

The simplest possible example that may help to understand the dynamics of vibrations is the oscillating point mass, which has one degree of freedom or single degree-of-freedom (SDOF) system. An SDOF system is one whose motion is governed by a single, second-order differential equation. Only two variables, position and velocity are needed to describe the trajectory of the system. Degree of freedom means the minimum number of independent coordinates required to determine completely the position of all parts of a system at any moment in time. An SDOF system is schematically illustrated in Figure 7.1. Let us assume that damping is negligible and there is no external force acting on the system. The vibrating mass, often referred to as the simple harmonic oscillator, has a mass of m and is sliding on a frictionless surface. The mass is connected to a surface with a linear spring, having the spring constant, k. Spring constant is also known as the stiffness of the system – the combined material and geometric property which resists the deformation of the body. According to Newton's Second Law of Motion, there is an inertia force, $F = ma = m\ddot{x}$, generated by the mass, which is proportional to its acceleration. Note that $\ddot{x}$ means the second derivative of x with respect to time (t), i.e., $\ddot{x} = \dfrac{d}{dt}\dfrac{d}{dt}(x) = \dfrac{d^2x}{dt^2}$. There is another force, $F = k(x + \delta_{st})$, acting against this, which is proportional to the spring constant k, where δ_{st} is the static/initial deflection of the system due to its own weight. The equation of motion may be expressed as:

$$m\ddot{x} = mg - k(x + \delta_{st})$$

where m is mass of the system, k is the spring constant of the system, and x is the displacement of the system from static equilibrium.

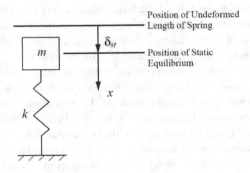

FIGURE 7.1   Single Degree-of-Freedom System

As the initial deformation (δ_{st}) occurs due to the weight of the body:

$$Weight = k\delta_{st}$$

$$mg = k\delta_{st}$$

Then, the equation of motion can be written as:

$$m\ddot{x} = mg - kx - k\delta_{st}$$

$$m\ddot{x} = mg - kx - mg$$

$$m\ddot{x} + kx = 0$$

or, $\ddot{x} + \left(\dfrac{k}{m}\right)x = 0$

This is a second-order differential equation. The solution procedure of second-order differential equation is described in Appendix A. Readers are encouraged to review Appendix A for better understanding. The solution of this second-order differential equation is:

$$x(t) = C_1 \cos(\omega_n t) + C_2 \sin(\omega_n t)$$

where $\omega_n = \sqrt{\dfrac{k}{m}}$ is the undamped natural angular frequency and C_1 and C_2 are constants of integration whose values are determined from the initial conditions. The angular frequency (ω_n) of an object undergoing periodic motion measures the rate at which a particle sweeps through a full 360 degrees, or 2π radians. It refers to the angular displacement per unit time and is expressed as rad/sec. If the initial conditions are denoted as $x(0) = x_o$ and $\dot{x}(0) = v_o$, then:

$$x(t) = x_o \cos(\omega_n t) + \left(\dfrac{v_o}{\omega_n}\right) \sin(\omega_n t)$$

As $mg = k\delta_{st}$, $\dfrac{k}{m} = \dfrac{g}{\delta_{st}}$, the undamped natural angular frequency (ω_n) may be expressed in terms of the static deflection of the system as:

$$\omega_n = \sqrt{\dfrac{g}{\delta_{st}}}$$

The undamped natural time period of vibration (τ_n) may now be written as:

$$\tau_n = \dfrac{2\pi}{\omega_n} = \dfrac{2\pi}{\sqrt{\dfrac{k}{m}}} = \dfrac{2\pi}{\sqrt{\dfrac{g}{\delta_{st}}}}$$

Recall that the time period of vibration is the time taken for one complete cycle of vibration to pass a given point. The unit of time period is sec. As the frequency of a wave increases, the time period of the wave decreases. Frequency (f) and time period (τ) are in a reciprocal relationship that can be expressed mathematically as:

$$\tau = \frac{1}{f}$$

$$f = \frac{1}{\tau}$$

The unit of linear frequency is per sec or sec^{-1} or Hz. Linear frequency (f) counts the number of complete oscillations or rotations in a given period of time. Angular frequency (ω) measures angular displacement per unit time and is expressed as rad/sec. Wavelength (λ) is the distance between identical points (adjacent crests) in the adjacent cycles of a waveform signal propagated in space or along a wire. In wireless systems, this length is usually specified in m or ft.

Example 7.1

A suspension system of a car is designed for a 2,000 kg vehicle (empty weight), as shown in Figure 7.2. It is estimated that the maximum added mass from passengers and cargo is 1,000 kg. When the vehicle is empty, its static deflection is 3.1 mm. Calculate the spring constant of the suspension system.

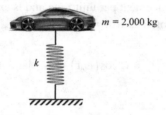

$m = 2,000$ kg

k

FIGURE 7.2   A Suspension System of a Car for Example 7.1

SOLUTION

Given,

Empty mass, $m = 2,000$ kg (the static deflection is given with respect to its empty mass)
Static deflection, $\delta_{st} = 3.1$ mm
Spring constant, $k = ?$
Known: $mg = k\delta_{st}$

Therefore, $k = \dfrac{mg}{\delta_{st}} = \dfrac{(2,000\,\text{kg})\left(9.81\dfrac{\text{m}}{\text{sec}^2}\right)}{0.0031\text{m}} = 6.33\times10^{6}\,\dfrac{\text{N}}{\text{m}}$

Answer: 6.33 MN/m

Example 7.2

A 200-kg machine is placed at the end of a 1.8-m-long steel cantilever beam, as shown in Figure 7.3. The machine is observed to vibrate with a natural frequency of 21 Hz. Calculate the undamped natural time period of vibration.

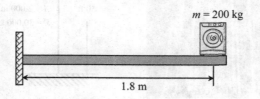

FIGURE 7.3 A Steel Cantilever Beam for Example 7.2

SOLUTION

The natural frequency of the system, $\omega_n = \left(21\dfrac{\text{cycle}}{\text{sec}}\right)\left(2\lambda\dfrac{\text{rad}}{\text{cycle}}\right)$

$$=131.9\dfrac{\text{rad}}{\text{sec}}$$

Natural time period, $\tau_n = \dfrac{2\pi}{\omega_n}$

$$=\dfrac{2\pi}{131.9\dfrac{\text{rad}}{\text{sec}}}$$

$$=0.048\,\text{sec}$$

Answer: 0.048 sec

Example 7.3

Consider the South Boston Municipal Water Tower be assumed to be supported by a single column, as shown in Figure 7.4. The moment of inertia (I), cross-sectional area (A), and stiffness (k) are the combined effect of all columns. Ignoring the weight of the column, calculate the following:

(a) The natural angular frequency in rad/sec, frequency in Hz, and the period of vibration in sec along the horizontal direction. (Note: Stiffness, $k = 3EI/L^3$ in the horizontal/lateral direction)

(b) The natural angular frequency in rad/sec, frequency in Hz, and the period of vibration in sec along the vertical direction. (Note: Stiffness, $k = EA/L$ in the vertical/axial direction)

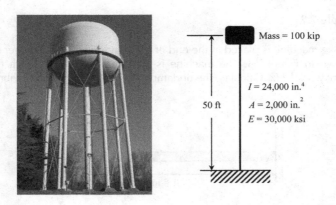

FIGURE 7.4 South Boston Municipal Water Tower, Virginia (Numerical Data Are Hypothetical)

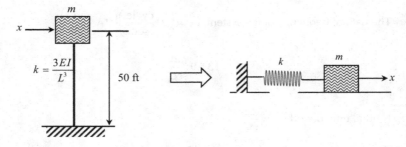

FIGURE 7.5 Analysis of the Water Tower along the Horizontal/Lateral Direction

SOLUTION

$$\text{Mass, } m = 100\,\text{kip} = \frac{100\,\text{kip}\left(1{,}000\dfrac{\text{lb}}{\text{kip}}\right)}{32.2\dfrac{\text{ft}}{\text{sec}^2}} = 3{,}105.6\,\text{slug (Figure 7.5)}$$

(a) Stiffness, $k = \dfrac{3EI}{L^3}$

$$= \frac{3\left(30{,}000\text{x}10^3\,\text{psi}\right)\left(24{,}000\,\text{in.}^4\right)}{\left(50\text{x}12\,\text{in.}\right)^3}$$

$$= 10{,}000\,\frac{\text{lb}}{\text{in.}}$$

$$= 120{,}000\,\frac{\text{lb}}{\text{ft}}$$

Natural angular frequency, $\omega_n = \sqrt{\dfrac{k}{m}}$

$$= \sqrt{\dfrac{120,000\dfrac{\text{lb}}{\text{ft}}}{3105.6\,\text{slug}}}$$

$$= 6.22\dfrac{\text{rad}}{\text{sec}}$$

Natural time period, $\tau_n = \dfrac{2\pi}{\omega_n} = \dfrac{2\pi}{6.22\dfrac{\text{rad}}{\text{sec}}} = 1.01\text{sec}$

Frequency, $f = \dfrac{1}{1.01\text{sec}} = 0.99\,\text{Hz}$

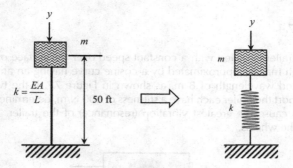

FIGURE 7.6 Analysis of the Water Tower along the Vertical/Axial Direction

(b) Vertical Direction (Figure 7.6)

Stiffness, $k = \dfrac{EA}{L}$

$$= \dfrac{\left(30,000\text{x}10^3\,\text{psi}\right)\left(2,000\,\text{in.}^2\right)}{\left(50\text{x}12\,\text{in.}\right)}$$

$$= 100,000,000\dfrac{\text{lb}}{\text{in.}}$$

$$= 1,200,000,000\dfrac{\text{lb}}{\text{ft}}$$

Natural angular frequency, $\omega_n = \sqrt{\dfrac{k}{m}}$

$$= \sqrt{\frac{1,200,000,000\frac{lb}{ft}}{3,105.6\,slug}}$$

$$= 621.6\frac{rad}{sec}$$

Natural time period, $\tau_n = \dfrac{2\pi}{\omega_n} = \dfrac{2\pi}{621.6\dfrac{rad}{sec}} = 0.01\,sec$

Frequency, $f = \dfrac{1}{0.01\,sec} = 100\,Hz$

Answers:

- *6.22 rad/sec, 0.99 Hz, and 1.01 sec*
- *621.6 rad/sec, 100 Hz, and 0.01 sec*

Example 7.4

A 450-kg trailer is pulled with a constant speed over the surface of a bumpy road, which may be approximated by a cosine curve having an amplitude of 100 mm and wavelength of 8 m, as shown in Figure 7.7. If the two springs which support the trailer each have a stiffness of 800 N/m, determine the speed which will cause the greatest vibration (resonance) of the trailer. Ignore the weight of the wheels.

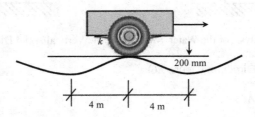

FIGURE 7.7   The Surface of a Bumpy Road for Example 7.4

SOLUTION

The amplitude is $\delta_o = 100$ mm $= 0.1$ m
The wavelength, $\lambda = 8$ m
Combined stiffness, $k = 2\,(800\ N/m) = 1,600\ N/m$

Natural angular frequency, $\omega_n = \sqrt{\dfrac{k}{m}} = \sqrt{\dfrac{1,600\dfrac{N}{m}}{450\ kg}} = 1.89\,rad$

Time period, $\tau_n = \dfrac{2\pi}{\omega_n} = \dfrac{2\pi}{1.89\,rad} = 3.33\,sec$

For maximum vibration of the trailer, resonance must occur, i.e., $\omega_o = \omega_n$. Thus, the trailer must travel $\lambda = 8$ m in $\tau = 3.33$ sec so that

$$v_R = \frac{\lambda}{\tau} = \frac{8\,m}{3.33\,sec} = 2.40\,m/sec$$

Answer: 2.4 m/sec

7.3 UNDAMPED-FREE TORSIONAL VIBRATION

Torsional vibration is the angular vibration of an object – commonly a shaft along its axis of rotation, as shown in Figure 7.8. Torsional vibration is often a concern in power transmission systems using rotating shafts or couplings where it can cause failures if not controlled. A second effect of torsional vibration applies to passenger cars which can lead to seat vibrations or noise at certain speeds. Both reduce the comfort of the ride. In ideal power generation, or transmission systems using rotating parts, it is considered that the applied, or reacted torques are smooth and leads to constant speeds. and it is also considered that the rotating plane where the power is generated (or input), and the plane from where the power is taken out (output) are the same. In reality, these are not the case. The torques generated may not be smooth (e.g., internal combustion engines), or the component being driven may not react to the torque smoothly (e.g., reciprocating compressors), and the power generating plane is normally at some distance to the power takeoff plane. Also, the components transmitting the torque can generate non-smooth or alternating torques (e.g., elastic drive belts, worn gears, misaligned shafts). Because no material can be infinitely stiff, these alternating torques applied at some distance on a shaft cause twisting vibrations about the axis of rotation.

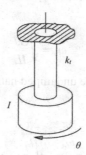

FIGURE 7.8   Schematic of Torsional Vibration

For torsional free vibrations, it may be shown that the differential equation of motion is:

$$\ddot{\theta} + \left(\frac{k_t}{I}\right)\theta = 0$$

where:

$\ddot{\theta}$ = the second derivative of the angular displacement (θ) of the system with respect to time (t), i.e., $\dfrac{d}{dt}\dfrac{d}{dt}(\theta) = \dfrac{d^2\theta}{dt^2}$.

k_t = the torsional stiffness of the massless rod
I = the mass moment of inertia of the end mass

The solution may now be written in terms of the initial conditions: $\theta\,(0) = \theta_o$ and $\dot{\theta}(0) = \dot{\theta}_o$

$$\theta(t) = \theta_o \cos(\omega_n t) + \left(\dfrac{\dot{\theta}_o}{\omega_n}\right)\sin(\omega_n t)$$

where $\dot{\theta}_o$ means the derivative of θ with respect to time (t), i.e., $\dot{\theta}_o = \dfrac{d}{dt}(\theta_o)$. The undamped natural angular frequency is given by:

$$\omega_n = \sqrt{\dfrac{k_t}{I}}$$

The torsional stiffness of a solid round rod with associated polar moment of inertia J, length L, and shear modulus of elasticity (also called the modulus of rigidity) G is given by:

$$k_t = \dfrac{GJ}{L}$$

Thus, the undamped natural angular frequency for a system with a solid-round-supporting rod can be written as:

$$\omega_n = \sqrt{\dfrac{GJ}{IL}}$$

Similar to the linear vibration, the undamped natural period (ω_n) may be written as:

$$\tau_n = \dfrac{2\pi}{\omega_n} = \dfrac{2\pi}{\sqrt{\dfrac{k_t}{I}}} = \dfrac{2\pi}{\sqrt{\dfrac{GJ}{IL}}}$$

Example 7.5

A torsional pendulum consists of a 10 kg uniform circular disk with a radius of 0.5 m attached at its center to a weightless 3-m-long rod, as shown in Figure 7.9. Calculate the undamped natural angular frequency of the pendulum if the torsional spring constant is 0.5 N.m/rad.

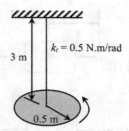

FIGURE 7.9   A Torsional Pendulum for Example 7.5

SOLUTION

Given,

Empty mass, $m = 10$ kg
Radius, $R = 0.5$ m
Torsional spring constant, $k_t = 0.5$ N.m/rad
Undamped natural angular frequency, $\omega_n = ?$

Mass moment of inertia of the circular disk, $I = \dfrac{MR^2}{2}$

$$= \frac{(10\,\text{kg})(0.5\,\text{m})^2}{2}$$

$$= 1.25 \text{ kg.m}^2$$

Natural angular frequency, $\omega_n = \sqrt{\dfrac{k_t}{I}} = \sqrt{\dfrac{0.5\dfrac{\text{N.m}}{\text{rad}}}{1.25\,\text{kg.m}^2}} = 0.63\,\dfrac{\text{rad}}{\text{sec}}$

Answer: 0.63 rad/sec

7.4   UNDAMPED FORCED VIBRATION

Undamped forced vibration is considered to be one of the most important types of vibrating motion in engineering. Its principles can be used to describe the motion of many types of machines and structures. Most engineering structures fall into this category.

7.4.1   PERIODIC FORCE

The block and spring shown in Figure 7.10a provide a convenient model which represents the vibrational characteristics of a system subjected to a periodic force $F = F_o \sin(w_o t)$. This force has an amplitude of F_o and a forcing angular frequency

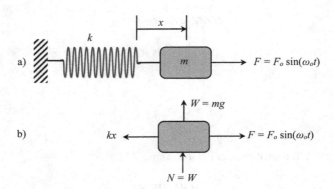

FIGURE 7.10   A Spring System with a Periodic Force

w_o. The free-body diagram for the block when it is displaced a distance x is shown in Figure 7.10b. Applying the equation of motion, we have

$$\sum F_x = ma_x$$

$$F_o \sin(\omega_o t) - kx = m\ddot{x}$$

$$\frac{F_o}{m} \sin(\omega_o t) - \frac{k}{m} x = \frac{m}{m} \ddot{x}$$

$$\ddot{x} + \frac{k}{m} x = \frac{F_o}{m} \sin(\omega_o t)$$

This equation is a nonhomogeneous second-order differential equation. The general solution consists of a complementary solution, x_c, plus a particular solution, x_p. The complementary solution is determined by setting the term on the right side of the equation equal to zero and solving the resulting homogeneous equation. The solution can be shown as:

$$x_c = A\sin(\omega_n t) + B\cos(\omega_n t)$$

Where ω_n is the natural frequency, $\omega_n = \sqrt{\dfrac{k}{m}}$ and A and B are constants. Since the motion is periodic, the particular solution can be determined by assuming a solution of the form:

$$x_p = X\sin(\omega_o t)$$

where X is a constant. Taking the second-time derivative of $x_p = X\sin(\omega_o t)$ and substituting into $\ddot{x} + \dfrac{k}{m}x = \dfrac{F_o}{m}\sin(\omega_o t)$ yields:

$$-X\omega_o^2 \sin(\omega_o t) + \frac{k}{m}\big(X\sin(\omega_o t)\big) = \frac{F_o}{m}\sin(\omega_o t)$$

Factoring out $\sin\omega_o t$ and solving for X gives:

$$X = \frac{\dfrac{F_o}{m}}{\left(\dfrac{k}{m}\right) - \omega_o^2} = \frac{\dfrac{F_o}{k}}{1 - \left(\dfrac{\omega_o}{\omega_n}\right)^2}$$

Then, the particular solution becomes:

$$x_p = \frac{\dfrac{F_o}{k}}{1 - \left(\dfrac{\omega_o}{\omega_n}\right)^2}\sin(\omega_o t)$$

The general solution is therefore the sum of two sine functions having different frequencies.

$$x = x_c + x_p = A\sin(\omega_n t) + B\cos(\omega_n t) + \frac{\dfrac{F_o}{k}}{1 - \left(\dfrac{\omega_o}{\omega_n}\right)^2}\sin(\omega_o t)$$

The complementary solution x_c defines the free vibration, which depends on the natural frequency $\omega_n = \sqrt{\dfrac{k}{m}}$ and the constants C and ϕ. The particular solution, x_p describes the forced vibration of the block caused by the applied force $F = F_o\sin(\omega_o t)$. Since all vibrating systems are subject to friction, the free vibration, x_c, will in time dampen out and become small in magnitude, as shown in Figure 7.11. For this reason, the free vibration (x_c) is referred to as transient, and the forced vibration (x_p) is called steady state, since it is the only vibration that remains. Free vibration (x_c) is very often neglected during analysis. It can be seen that the amplitude of forced or steady-state vibration depends on the frequency ratio $\dfrac{\omega_o}{\omega_n}$. If the magnification factor, MF, is defined as the ratio of the amplitude of steady-state vibration, X, to the static deflection, F_o/k, which would be produced by the amplitude of the periodic force F_o, then:

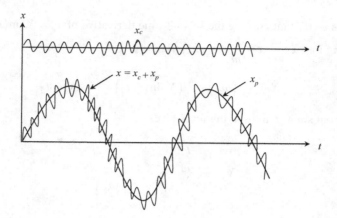

FIGURE 7.11 Complementary, Particular, and General Solutions of Undamped Forced Vibration for Periodic Load

$$MF = \frac{X}{\dfrac{F_o}{k}} = \frac{1}{1 - \left(\dfrac{\omega_o}{\omega_n}\right)^2}$$

If the force or displacement is applied with a frequency close to the natural frequency of the system, i.e., $\dfrac{\omega_o}{\omega_n} \approx 1$, the amplitude of vibration of the block becomes extremely large. This occurs because the force F is applied to the block so that it always follows the motion of the block. This condition is called resonance, and in practice, resonating vibrations can cause tremendous stress and rapid failure of parts.

Example 7.6

Figure 7.12 shows a single-story moment frame consisting of two steel columns and a rigid diaphragm. The total dead load on the diaphragm including all roof supporting members is 100 kip. The elastic modulus of column $E = 29,000$ ksi. The columns are 15 ft tall with the moment of inertia of each column, $I = 2,000$ in.⁴, and are fixed at the foundation. There is no damping used in the frame. The frame is subjected to an external horizontal vibration force, $F = 150\sin(20t)$ kip, where t is in sec. Stiffness of each column can be calculated as $k = \dfrac{12EI}{L^3}$.

Determine the following parameters of the moment frame for the vibration along the horizontal direction:

(a) Natural angular frequency, ω_n
(b) Natural time period, τ_n
(c) Frequency of vibration, f

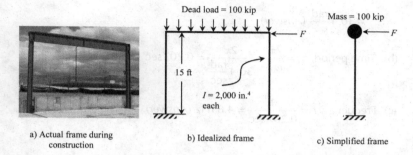

a) Actual frame during construction

b) Idealized frame

c) Simplified frame

FIGURE 7.12 A Single-Story Moment Frame for Example 7.6

(d) The displacement of the forced vibration of the block caused by the applied force, i.e., 'x_p' and the maximum displacement of the steady-state vibration, i.e., 'X'.

(e) The amplitude of the velocity of the steady-state vibration i.e., 'v_p' and the maximum velocity

(f) The amplitude of the acceleration of the steady-state vibration i.e., 'a_p' and the maximum acceleration

SOLUTION

$$\text{Mass, } m = 100\,\text{kip} = \frac{100\,\text{kip}\left(1{,}000\dfrac{\text{lb}}{\text{kip}}\right)}{32.2\dfrac{\text{ft}}{\text{sec}^2}} = 3{,}105.6\,\text{slug}$$

$$\text{Combined stiffness, } \sum k = 2.\frac{12EI}{L^3}$$

$$= \frac{24\left(29{,}000{,}000\,\text{psi}\right)\left(2{,}000\,\text{in.}^4\right)}{\left(15 \times 12\ \text{in.}\right)^3}$$

$$= 238{,}683\,\frac{\text{lb}}{\text{in.}}$$

$$= 2{,}864{,}198\,\frac{\text{lb}}{\text{ft}}$$

(a) Natural angular frequency, $\omega_n = \sqrt{\dfrac{k}{m}}$

$$= \sqrt{\frac{2{,}864{,}198\dfrac{\text{lb}}{\text{ft}}}{3105.6\,\text{slug}}}$$

$$= 30.4 \frac{rad}{sec} \ (Answer)$$

(b) Time period, $\tau_n = \dfrac{2\pi}{\omega_n} = \dfrac{2\pi}{30.4 \dfrac{rad}{sec}} = 0.207 \, sec \ (Answer)$

(c) Frequency, $f = \dfrac{1}{0.207 \, sec} = 4.83 \, Hz \ (Answer)$

(d) $x_p = \dfrac{\dfrac{F_o}{k}}{1 - \left(\dfrac{\omega_o}{\omega_n}\right)^2} \sin \omega_o t$

$$= \frac{\dfrac{150,000}{2,864,198}}{1 - \left(\dfrac{20}{30.4}\right)^2} \sin(20t)$$

$$= 0.0923 \sin(20t) \ (Answer)$$

Maximum displacement $= 0.0923$ ft $(Answer)$

(e) The amplitude of the velocity of the steady-state vibration

$$v_p = \frac{dx_p}{dt}$$

$$= \frac{d}{dt}\big(0.0923 \sin(20t)\big)$$

$$= 0.0923(20)\cos(20t)$$

$$= 1.846 \cos(20t) \ (Answer)$$

Maximum velocity $= 1.846$ ft/sec $(Answer)$

(f) The amplitude of the acceleration of the steady-state vibration

$$a_p = \frac{dv_p}{dt}$$

$$= \frac{d}{dt}\big(1.846 \cos(20t)\big)$$

$$= 1.846(20)\big(-\sin(20t)\big)$$

$$= -36.92 \sin(20t) \ (Answer)$$

Maximum acceleration $= 36.92$ ft/sec^2 $(Answer)$

7.4.2 Periodic Support Displacement

Forced vibrations can also arise from the periodic excitation of the support of a system. The model shown in Figure 7.13a represents the periodic vibration of a block which is caused by harmonic movement $\delta = \delta_o \sin(\omega_o t)$ of the support. The free-body diagram for the block in this case is shown in Figure 7.13b. The displacement δ of the support is measured from the point of zero displacement, i.e., when the radial line OA coincides with OB. Therefore, general deformation of the spring is $\left(x - \delta_o \sin(\omega_o t)\right)$. Applying the equation of motion yields:

$$\sum F_x = ma_x$$

$$-k\left(x - \delta_o \sin(\omega_o t)\right) = m\ddot{x}$$

$$-\frac{k}{m}\left(x - \delta_o \sin(\omega_o t)\right) = \frac{m}{m}\ddot{x}$$

$$\ddot{x} + \frac{k}{m}x = \frac{k\delta_o}{m}\sin(\omega_o t)$$

Now, $F_o = k\delta_o$

Then, $\ddot{x} + \dfrac{k}{m}x = \dfrac{F_o}{m}\sin(\omega_o t)$

This is identical to the equation obtained for periodic force in the previous subsection. Therefore, the results of periodic force are appropriate for describing the motion of the block when subjected to the support displacement, $\delta = \delta_o \sin(\omega_o t)$.

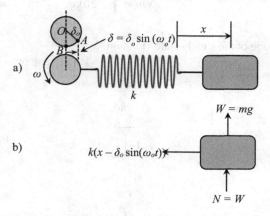

FIGURE 7.13 Vibration of an Undamped Forced System with a Periodic Support Displacement

Example 7.7

The instrument shown in Figure 7.14 is rigidly attached to a platform P, which in turn is supported by four springs, each having a stiffness $k = 800$ N/m. The floor is subjected to a vertical displacement $\delta = 10 \sin(8t)$ mm, where t is in sec. The instrument and platform have a total mass of 20 kg.

(a) Determine the amplitude of steady-state vibration.
(b) Calculate the frequency of the floor vibration required to cause resonance.

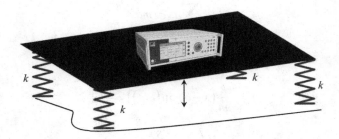

FIGURE 7.14 An Undamped Forced System with a Periodic Support Displacement for Example 7.7

SOLUTION

Periodic Support Displacement type load is applied.
$\delta = 10 \sin(8t)$ mm
$\delta_o = 10$ mm

Natural angular frequency, $\omega_n = \sqrt{\dfrac{k}{m}} = \sqrt{\dfrac{4\left(800\,\dfrac{N}{m}\right)}{20\,kg}} = 12.62\,\dfrac{rad}{sec}$

(a) Amplitude of the steady-state vibration, $X = \dfrac{\dfrac{F_o}{k}}{1-\left(\dfrac{\omega_o}{\omega_n}\right)^2}$

$$= \frac{\dfrac{k\delta_o}{k}}{1-\left(\dfrac{\omega_o}{\omega_n}\right)^2}$$

$$= \frac{\delta_o}{1-\left(\dfrac{\omega_o}{\omega_n}\right)^2}$$

$$= \cfrac{10}{1 - \left(\cfrac{8 \cfrac{rad}{sec}}{12.65 \cfrac{rad}{sec}} \right)^2}$$

$$=16.7\,mm$$

(b) Resonance occurs when the amplitude of vibration X caused by the floor displacement approaches infinity. This requires:

$$\frac{\omega_o}{\omega_n} \approx 1$$

$$\omega_o \approx \omega_n \approx 12.65 \frac{rad}{sec}$$

Answers:

Amplitude of the steady-state vibration = 16.7 mm
Frequency of the floor vibration required to cause resonance = 12.65 rad/sec

7.5 VISCOUS DAMPED FREE VIBRATION

The vibration analysis considered thus far has not included the effects of friction or damping in the system, and as a result, the solutions obtained are only in close agreement with the actual motion. Since all vibrations die out with time, the presence of damping forces should be included in the analysis. In many cases damping is attributed to the resistance created by the substance, such as water, oil, or air, in which the system vibrates. Provided the body moves slowly through this substance, the resistance to motion is directly proportional to the body's speed. The type of force developed under these conditions is called a viscous damping force. The magnitude of this force is expressed by an equation of the form:

$$F = c\dot{x}$$

where the constant c is called the coefficient of viscous damping and has units of N.sec/m or lb.sec/ft.

The vibrating motion of a body or system having viscous damping can be characterized by the block and spring shown in Figure 7.15a. The effect of damping is provided by the dashpot connected to the block on the right-hand side. Damping occurs when the piston P moves to the right or left within the enclosed cylinder. The cylinder contains a fluid, and the motion of the piston is retarded since the fluid must flow around or through a small hole in the piston. The dashpot is assumed to have a coefficient of viscous damping c. If the block is displaced a

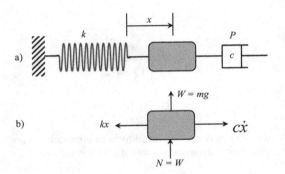

FIGURE 7.15   The Model of a Viscous Damped Free Vibration

distance x from its equilibrium position, the resulting free-body diagram is shown in Figure 7.15b.

Both the spring and damping force oppose the forward motion of the block, so that applying the equation of motion yields:

$$\sum F_x = ma_x$$

$$-kx - c\dot{x} = m\ddot{x}$$

$$m\ddot{x} + c\dot{x} + kx = 0$$

This linear, second-order, homogeneous, differential equation has a solution of the form:

$$x = e^{\lambda t}$$

where e is the base of the natural logarithm and λ (lambda) is a constant. The value of λ can be obtained by substituting this solution and its time derivatives into $m\ddot{x} + c\dot{x} + kx = 0$, which yields:

$$m\lambda^2 e^{\lambda t} + c\lambda e^{\lambda t} + ke^{\lambda t} = 0$$

$$e^{\lambda t}\left(m\lambda^2 + c\lambda + k\right) = 0$$

Since $e^{\lambda t}$ can never be zero, a solution is possible provided that:

$$m\lambda^2 + c\lambda + k = 0$$

Hence, by the quadratic formula, the two values of λ :

$$\lambda_1 = -\frac{c}{2m} + \sqrt{\left(\frac{c}{2m}\right)^2 - \frac{k}{m}}$$

$$\lambda_2 = -\frac{c}{2m} - \sqrt{\left(\frac{c}{2m}\right)^2 - \frac{k}{m}}$$

The general solution of $m\ddot{x} + c\dot{x} + kx = 0$ is therefore a combination of exponentials which involves both of these roots. There are three possible combinations of λ_1 and λ_2 which must be considered. Before discussing these combinations, however, we will first define the critical damping coefficient c_c as the value of c which makes the radical in the above two equations $\left(\left(\frac{c_c}{2m}\right)^2 - \frac{k}{m} = 0\right)$ equal to zero, i.e.,

$$\left(\frac{c_c}{2m}\right)^2 - \frac{k}{m} = 0$$

$$c_c = 2m\sqrt{\frac{k}{m}} = 2m\omega_n$$

Overdamped System. When $c > c_c$ the roots λ_1 and λ_2 are both real. The general solution of $m\ddot{x} + c\dot{x} + kx = 0$ can then be written as:

$$x = Ae^{\lambda_1 t} + Be^{\lambda_2 t}$$

Motion corresponding to this solution is nonvibrating. The effect of damping is so strong that when the block is displaced and released, it simply creeps back to its original position without oscillating. The system is said to be overdamped.

Critically Damped System. If $c = c_c$ then $\lambda_1 = \lambda_2 = -\frac{c_c}{2m} = -\omega_n$

This situation is known as critical damping, since it represents a condition where c has the smallest value necessary to cause the system to be nonvibrating. Using the methods of differential equations, it can be shown that the solution to $m\ddot{x} + c\dot{x} + kx = 0$ for critical damping is

$$x = (A + Bt)e^{-\omega_n t}$$

Underdamped System. Most often $c < c_c$ in which case the system is referred to as underdamped. In this case the roots λ_1 and λ_2 are complex numbers, and it can be shown that the general solution of $m\ddot{x} + c\dot{x} + kx = 0$ can be written as:

$$x = D\left[e^{-\left(\frac{c}{2m}\right)t}\sin(\omega_d t + \phi)\right]$$

where D and ϕ are constants generally determined from the initial conditions of the problem. The constant ω_d is called the damped natural frequency of the system. It has a value of

$$\omega_d = \sqrt{\frac{k}{m} - \left(\frac{c}{2m}\right)^2} = \omega_n\sqrt{1 - \left(\frac{c}{c_c}\right)^2}$$

where the ratio c/c_c is called the damping factor.

The initial limit of motion, D, diminishes with each cycle of vibration, since motion is confined within the bounds of the exponential curve. Using the damped natural frequency ω_d, the period of damped vibration can be written as:

$$\tau_d = \frac{2\pi}{\omega_d}$$

Since $\omega_d < \omega_n$, the period of damped vibration, τ_d, will be greater than that of free vibration, $\tau_n = \dfrac{2\pi}{\omega_n}$

Example 7.8

A block having a mass of 7 kg is suspended from a spring that has a stiffness of 600 N/m. Assume that positive displacement of the block is downward, and that motion takes place in a medium which furnishes a damping force $F = 50|v|$ N, where v is in m/sec. If the block is given an upward velocity from its equilibrium position at $t = 0$, determine the damped natural frequency of the system. Is the system underdamped or overdamped?

SOLUTION

Damping force, $F = 50|v|$. Therefore, the coefficient of viscous damping,
    $c = 50$ N.sec/m
Spring constant or stiffness, $k = 600$ N/m
Mass, $m = 7$ kg

Natural angular frequency, $\omega_n = \sqrt{\dfrac{k}{m}} = \sqrt{\dfrac{600\dfrac{N}{m}}{7\,kg}} = 9.258\,\dfrac{rad}{sec}$

Critical damping coefficient,

$$c_c = 2m\omega_n = 2(7\ kg)\left(9.258\frac{rad}{sec}\right) = 129.6\,\frac{N.sec}{m}$$

The damped natural frequency, $\omega_d = \omega_n\sqrt{1 - \left(\dfrac{c}{c_c}\right)^2}$

$$= 9.258 \frac{rad}{sec} \sqrt{1 - \left(\frac{50}{129.6}\right)^2}$$

$$= 8.542 \frac{rad}{sec}$$

As $c < c_c$, the system is underdamped.
Answers:

The system is underdamped.
The damped natural frequency = 8.542 rad/sec

7.6 VISCOUS DAMPED FORCED VIBRATION

The most general case of single degree-of-freedom vibrating motion occurs when the system includes the effects of forced motion and induced damping. The analysis of this particular type of vibration is of practical value when applied to systems having significant damping characteristics. If a dashpot is attached to the block and spring shown in Figure 7.16, the differential equation which describes the motion becomes:

$$m\ddot{x} + c\dot{x} + kx = F_o \sin(\omega_o t)$$

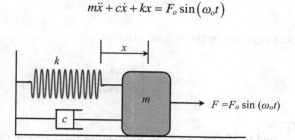

FIGURE 7.16 A System with both Spring (Elastic) and Dashpot (Damping)

Since the equation, $m\ddot{x} + c\dot{x} + kx = F_o \sin(\omega_o t)$ is nonhomogeneous, the general solution is the sum of a complementary solution, x_c, and a particular solution, x_p. The complementary solution is determined by setting the right side of $m\ddot{x} + c\dot{x} + kx = F_o \sin(\omega_o t)$ equal to zero and solving the homogeneous equation, which is equivalent to $m\ddot{x} + c\dot{x} + kx = 0$. The solution is therefore given in the previous section, depending on the values of λ_1 and λ_2. Because all systems are subjected to friction, this solution will dampen out with time. Only the particular solution, which describes the steady-state vibration of the system, will remain. Since the applied forcing function is harmonic, the steady-state motion will also be harmonic. Consequently, the particular solution will be of the form:

$$X_P = X' \sin(\omega_o t - \phi')$$

The constants X' and ϕ' are determined by taking the first- and second-time derivatives and substituting them into $m\ddot{x} + c\dot{x} + kx = F_o \sin(\omega_o t)$, which after simplification yields:

$$-X'm\omega_o^2 \sin(\omega_o t - \phi') + X'c\omega_o \cos(\omega_o t - \phi') + X'k \sin(\omega_o t - \phi') = F_o \sin(\omega_o t)$$

Since this equation holds for all time, the constant coefficients can be obtained by setting $(\omega_o t - \phi') = 0$ and $(\omega_o t - \phi') = \dfrac{\pi}{2}$ which causes the above equation to become

$$X'c\omega_o = F_o \sin\phi'$$

$$-X'm\omega_o^2 + X'k = F_o \cos\phi'$$

The amplitude as obtained by squaring these equations, adding the results, and using the identity $\sin^2\phi' + \cos^2\phi' = 1$ gives

$$X' = \frac{F_o}{\sqrt{\left(k - m\omega_o^2\right)^2 + c^2\omega_o^2}}$$

Dividing the first equation by the second gives:

$$\phi' = \tan^{-1}\left[\frac{c\omega_o}{k - m\omega_o^2}\right]$$

Since $\omega_n = \sqrt{\dfrac{k}{m}}$ and $c_c = 2m\omega_n$ then the above equations can also be written as:

$$X' = \frac{\dfrac{F_o}{k}}{\sqrt{\left[1 - \left(\dfrac{\omega_o}{\omega_n}\right)^2\right]^2 + \left[2\left(\dfrac{c}{c_c}\right)\left(\dfrac{\omega_o}{\omega_n}\right)\right]^2}}$$

$$\phi' = \tan^{-1}\left[\frac{2\left(\dfrac{c}{c_c}\right)\left(\dfrac{\omega_o}{\omega_n}\right)}{1 - \left(\dfrac{\omega_o}{\omega_n}\right)^2}\right]$$

The angle ϕ' represents the phase difference between the applied force and the resulting steady-state vibration of the damped system. The magnification factor

(MF) has been defined as the ratio of the amplitude of deflection caused by the forced vibration to the deflection caused by a static force F_o. Thus,

$$MF = \frac{X'}{\dfrac{F_o}{k}} = \cfrac{1}{\sqrt{\left[1 - \left(\dfrac{\omega_o}{\omega_n}\right)^2\right]^2 + \left[2\left(\dfrac{c}{c_c}\right)\left(\dfrac{\omega_o}{\omega_n}\right)\right]^2}}$$

The magnification of the amplitude increases as the damping factor decreases, as shown in Figure 7.17. Resonance obviously occurs only when the damping factor is zero and the frequency ratio equals 1.

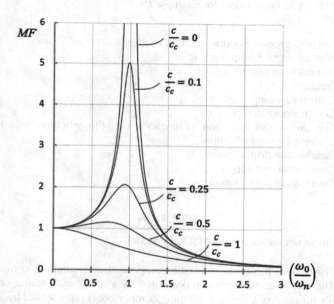

FIGURE 7.17 Variation of Magnification Factors Associated with Different Levels of (Under)damping for Viscous Damped Forced Vibration

Example 7.9

A moment frame has two steel columns and a rigid diaphragm, as shown in Figure 7.18. The total dead load including the weight of the frame is 50 kip. The lateral seismic force at the roof level can be expressed as $F = 20 \sin(5t)$ kip. The elastic modulus of the column is 29,000 ksi and the moment of inertia of each column is 800 in.[4]. The system has a damping with a ratio of 3%. One column is pinned to the footing ($k = 3EI/L^3$), while the other column is fixed to the footing ($k = 12EI/L^3$).

Determine the following of the moment frame for the vibration in the horizontal direction:

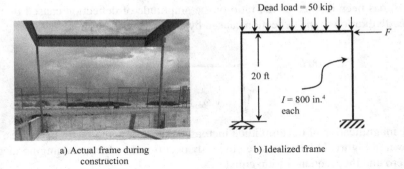

a) Actual frame during
construction

b) Idealized frame

FIGURE 7.18 A Moment Frame for Example 7.9

(a) Single degree-of-freedom (SDOF) model
(b) Natural angular frequency, ω_n
(c) Natural time period, τ_n
(d) Frequency, f
(e) Critical damping coefficient, c_c
(f) Coefficient of damping, c
(g) The differential equation of motion, i.e., the vibration equation (Damped forced vibration)
(h) Steady-state displacement, X_p
(i) Steady-state velocity, v_p
(j) Steady-state acceleration, a_p

SOLUTION

(a) Single degree-of-freedom (SDOF) model

The system has both an elastic component and damping component. The elastic component is represented by a spring with spring constant (or stiffness) of k and the damping component is represented by a dashpot with the damping coefficient of c. The SDOF can be presented, as shown in Figure 7.19. In this figure, m is the mass of the system and x is any horizontal displacement upon applying the periodic load $F = 20 \sin (5t)$.

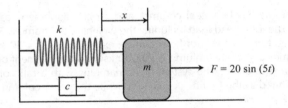

FIGURE 7.19 Analysis of the Moment Frame for Example 7.9

(b) Natural angular frequency, ω_n

We know the natural angular frequency, $\omega_n = \sqrt{\dfrac{k}{m}}$. First, let us find out the combined stiffness and the mass first before calculating the natural frequency.

Stiffness of Column 1, $k_1 = \dfrac{3EI}{L^3} = \dfrac{3(29,000,000\,\text{psi})(800\,\text{in.}^4)}{(20\text{x}12\,\text{in.})^3}$

$= 5,035\dfrac{\text{lb}}{\text{in.}} = 60,420\dfrac{\text{lb}}{\text{ft}}$

Stiffness of Column 2, $k_2 = \dfrac{12EI}{L^3} = \dfrac{12(29,000,000\,\text{psi})(800\,\text{in.}^4)}{(20\text{x}12\,\text{in.})^3}$

$= 20,139\dfrac{\text{lb}}{\text{in.}} = 241,667\dfrac{\text{lb}}{\text{ft}}$

Combined stiffness, $k = k_1 + k_2$
= 60,420 lb/ft + 241,667 lb/ft
= 302,087 lb/ft

Mass, $m = 50\,\text{kip} = \dfrac{50\,\text{kip}\left(1,000\dfrac{\text{lb}}{\text{kip}}\right)}{32.2\dfrac{\text{ft}}{\text{sec}^2}} = 1,552.8\,\text{slug}$

Now, $\omega_n = \sqrt{\dfrac{k}{m}} = \sqrt{\dfrac{302,087\dfrac{\text{lb}}{\text{ft}}}{1,552.8\,\text{slug}}} = 13.95\dfrac{\text{rad}}{\text{sec}}$ (Answer)

(c) Natural time period, τ_n

Time period, $\tau_n = \dfrac{2\pi}{\omega_n} = \dfrac{2\pi}{13.95\dfrac{\text{rad}}{\text{sec}}} = 0.45\,\text{sec}$ (Answer)

(d) Frequency, f

Frequency, $f = \dfrac{1}{0.45\,\text{sec}} = 2.22\,\text{Hz}$ (Answer)

(e) Critical damping coefficient, c_c

$c_c = 2m\omega_n = 2(1,552.8\,\text{slug})\left(13.95\dfrac{\text{rad}}{\text{sec}}\right) = 43,323.1\dfrac{\text{lb.sec}}{\text{ft}}$ (Answer)

(f) Coefficient of damping, c

$c = \left(\dfrac{c}{c_c}\right)c_c = (3\%)\left(43,323.1\dfrac{\text{lb.sec}}{\text{ft}}\right) = 1,300\dfrac{\text{lb.sec}}{\text{ft}}$ (Answer)

(g) The differential equation of motion, i.e., the vibration equation (Damped forced vibration)

$$m\ddot{x} + c\dot{x} + kx = F_o \sin\omega_o t$$

$$1,552.8\ddot{x} + 1,300\dot{x} + 302,087x = 20,000\sin(5t)$$

where m is in slug, c is in lb.sec/ft, k is in lb/ft, F is in lb, and ω_o is in rad/sec. The above equation can also be written as:

$$1,552.8\frac{d^2x}{dt^2} + 1,300\frac{dx}{dt} + 302,087x = 20,000\sin(5t) \quad (Answer)$$

(h) Steady-state displacement, X_p

$$X_p = X'\sin(\omega_o t - \phi')$$

Where, $X' = \dfrac{\dfrac{F_o}{k}}{\sqrt{\left[1-\left(\dfrac{\omega_o}{\omega_n}\right)^2\right]^2 + \left[2\left(\dfrac{c}{c_c}\right)\left(\dfrac{\omega_o}{\omega_n}\right)\right]^2}}$

$$X_p = \dfrac{\dfrac{F_o}{k}}{\sqrt{\left[1-\left(\dfrac{\omega_o}{\omega_n}\right)^2\right]^2 + \left[2\left(\dfrac{c}{c_c}\right)\left(\dfrac{\omega_o}{\omega_n}\right)\right]^2}}\sin(\omega_o t - \phi')$$

$$= \dfrac{\dfrac{20,000\,\text{lb}}{302,087\,\dfrac{\text{lb}}{\text{ft}}}}{\sqrt{\left[1-\left(\dfrac{5}{13.95}\right)^2\right]^2 + \left[2(0.03)\left(\dfrac{5}{13.95}\right)\right]^2}}\sin(\omega_o t - \phi')$$

$$= \frac{0.0662\,\text{ft}}{0.8717}\sin(5t - \phi')$$

Now, $\phi' = \tan^{-1}\left[\dfrac{2\left(\dfrac{c}{c_c}\right)\left(\dfrac{\omega_o}{\omega_n}\right)}{1-\left(\dfrac{\omega_o}{\omega_n}\right)^2}\right]$

$$= \tan^{-1}\left[\frac{2(0.03)\left(\dfrac{5}{13.95}\right)}{1-\left(\dfrac{5}{13.95}\right)^2}\right]$$

$$= \tan^{-1}(0.02468)$$

$$= 1.414°$$

$$= 1.414°\left(\frac{\pi}{180°}\right)$$

$$= 0.0247\,\text{rad}$$

Finally, $X_p = \dfrac{0.0662}{0.8717}\sin(5t - 0.0247)$ ft (Answer)

(i) Steady-state velocity, v_p

$$X_p = \frac{0.0662}{0.8717}\sin(5t - 0.0247)\ \text{ft}$$

$$v_p = \frac{dX_p}{dt}$$

$$= \frac{d}{dt}\left(\frac{0.0662}{0.8717}\sin(5t - 0.0247)\right)$$

$$= \frac{0.0662}{0.8717}\frac{d}{dt}\left(\sin(5t - 0.0247)\right)$$

$$= 0.0759(5)\cos(5t - 0.0247)$$

$$= 0.3795\cos(5t - 0.0247)\ \frac{\text{ft}}{\text{sec}}\ (Answer)$$

(j) Steady-state acceleration, a_p

$$a_p = \frac{d(v_p)}{dt}$$

$$= \frac{d}{dt}\left(0.3795\cos(5t - 0.0247)\right)$$

$$= 0.3795(5)\left(-\sin(5t - 0.0247)\right)$$

$$= -1.8975\sin(5t - 0.0247)\ \frac{\text{ft}}{\text{sec}^2}\ (Answer)$$

FUNDAMENTALS OF ENGINEERING (FE) EXAM STYLE QUESTIONS

FE Problem 7.1

Which of the following is true for a system consisting of a mass oscillating on the end of an ideal spring?

    A. the kinetic and potential energies are equal to each other at all times
    B. the kinetic and potential energies are both constant
    C. the maximum potential energy is achieved when the mass passes through its equilibrium position
    D. the maximum kinetic energy and maximum potential energy are equal, but occur at different times

FE Problem 7.2

For a particle moving with a simple harmonic motion, the frequency is:

    A. directly proportional to periodic time
    B. inversely proportional to periodic time
    C. inversely proportional to its angular velocity
    D. directly proportional to its angular velocity
    E. none of these options

FE Problem 7.3

Two objects of equal mass hang from independent springs of unequal spring constant and oscillate up and down. The spring with greater spring constant (or spring stiffness) must have the:

    A. smaller amplitude of oscillation
    B. larger amplitude of oscillation
    C. shorter period of oscillation
    D. longer period of oscillation

FE Problem 7.4

A spring is compressed by a 400-kg block, as shown in Figure 7.20. If the block is displaced 40 mm downward from its equilibrium position and given a downward velocity of 0.75 m/sec, the natural angular frequency (rad/sec) of the system is most nearly:

    A. 5.22
    B. 9.68
    C. 15.66
    D. 23.21

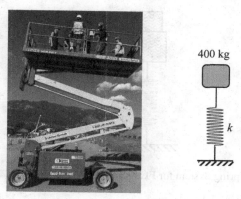

FIGURE 7.20   A Spring System for FE Problem 7.4

FE Problem 7.5

Periodic time of a particle moving with simple harmonic motion is the time taken by the particle for:

A. half oscillation
B. quarter oscillation
C. complete oscillation
D. none of these options

FE Problem 7.6

A particle moving with a simple harmonic motion, attains its maximum velocity when it passes:

A. the extreme points of the oscillation
B. through the mean/lowest position
C. through a point at half of the amplitude
D. none of these options

FE Problem 7.7

A 5-kilogram block is fastened to an ideal vertical spring that has an unknown spring constant. A 3-kilogram block rests on top of the 5-kilogram block, as shown in Figure 7.21. When the blocks are at rest, the spring is compressed to its equilibrium position a distance of 20 cm from its original length. The 3-kg block is then raised 50 cm above the 5-kg block and dropped onto it.

The resulting time period (sec) of this oscillation is most nearly:

A. 0.9
B. 1.3
C. 0.6
D. 1.8

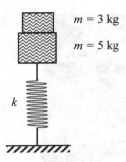

$m = 3$ kg

$m = 5$ kg

k

FIGURE 7.21   A Spring System for FE Problem 7.7

PRACTICE PROBLEMS

Section I: Undamped-Free Linear Vibration

Problem 7.1

A suspension system is being designed for a 3,500-kg vehicle (empty weight). It is estimated that the maximum added mass from passengers and cargo is 1,200 kg. When the vehicle is empty, its static deflection is to be 3.9 mm. Calculate the spring constant of the system.

Problem 7.2

The electric motor, shown in Figure 7.22, turns an eccentric flywheel which is equivalent to an unbalanced 0.25-lb weight located 10 in from the axis of rotation. If the static deflection of the beam is 1 in due to the weight of the motor, determine the angular velocity of the flywheel at which resonance will occur. The motor weighs 150 lb. Ignore the mass of the beam.

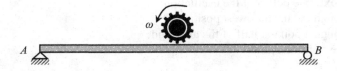

ω

A B

FIGURE 7.22   A Simply Supported Beam for Problem 7.2

Problem 7.3

When a 1,000-kg block is loaded on a lift, the lift is compressed a distance of 100 mm, as shown in Figure 7.23. Determine the natural frequency and the period of vibration for a 200-kg block attached to the same lift.

Section II: Undamped-Free Torsional Vibration

Problem 7.4

A torsional pendulum consists of a 10-kg uniform disk with a radius of 0.25 m attached at its center to a weightless 5-m-long rod. Calculate the undamped

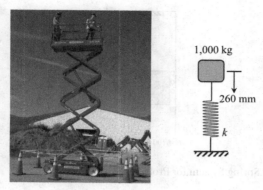

FIGURE 7.23 A Spring System for Problem 7.3

natural angular frequency of the pendulum if the torsional spring constant is 0.75 N.m/rad.

Problem 7.5

A torsional pendulum consists of an 18-kg uniform disk with a radius of 0.75 m attached at its center to a weightless 3.5-m-long rod. Calculate the undamped natural angular frequency of the pendulum if the torsional spring constant is 0.35 N.m/rad.

Problem 7.6

A 250-kg machine is placed at the end of a 1.6-m-long steel cantilever beam. The machine is observed to vibrate with a natural frequency of 25 Hz. Calculate the undamped natural period of vibration.

Problem 7.7

A 5-kg wheel is mounted on a 1-kg plate whose center is attached to a 40-mm diameter, 75-cm-long steel bar (G = 80 GPa), which is fixed at its other end. The centroidal polar moment of inertia of the plate is 1.4 kg.m². The period of torsional oscillation of this assembly is 0.15 sec. Calculate the polar moment of inertia of the wheel.

Section III: Undamped Forced Vibration

Problem 7.8

A 40-kg block is attached to a spring having a stiffness of 800 N/m, as shown in Figure 7.24. A force $F = (100 \cos(2t))$ N, where t is in seconds, is applied to the block. Determine the maximum speed of the block for the steady-state vibration.

Problem 7.9

A 30-kip cylindrical tank is attached with two cables having a stiffness of 5,000 lb/ft each, as shown in Figure 7.25. While lifting, a periodic force of $F = 8 \cos(3t)$ kip is applied, where t is in sec. Determine the maximum speed of the tank.

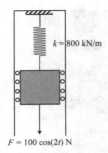

FIGURE 7.24   A Spring System for Problem 7.8

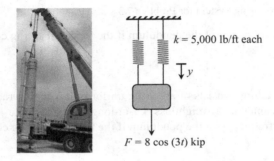

FIGURE 7.25   A Double-Spring System for Problem 7.9

Problem 7.10

While lifting a 5,000-kg concrete box a spring having a stiffness of 300 kN/m is used, as shown in Figure 7.26. If the block is acted upon by a vertical force $F = 7 \sin(8t)$ kN, where t is in seconds, determine the equation which describes the motion of the block when it is pulled down 100 mm from the equilibrium position and released from rest at $t = 0$. Assume that positive displacement is downward.

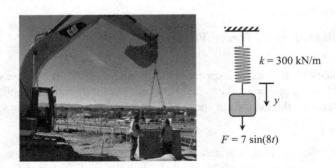

FIGURE 7.26   A Spring System for Problem 7.10

Section IV: Viscous Damped Free Vibration

Problem 7.11

A structure is idealized as a damped spring-mass system with stiffness 10 kN/m; mass 2 Mg; and dashpot coefficient 2 kN.sec/m. It is subjected to a harmonic force of amplitude 500 N at frequency 0.5 Hz. Calculate the steady state amplitude of vibration.

Problem 7.12

The mass-spring-dashpot system shown in Figure 7.27 consists of a mass of $m =$ 2 kg, a spring with stiffness of $k = 3,200$ N/m and a dashpot with damping coefficient of $c = 100$ N.sec/m.

    a. Is the system underdamped, critically damped, or overdamped?
    b. Find the damped natural frequency of the system.

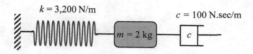

FIGURE 7.27   A Spring-Dashpot System for Problem 7.12

Section V: Viscous Damped Forced Vibration

Problem 7.13

Figure 7.28 shows a beam with a span of 25 ft. The total dead load including the weight of the beam is 25 kip and can be assumed concentrated at the mid-span. The elastic modulus of the column is 4,000 ksi and the moment of inertia of the beam section is 15,000 in.[4]. The system has a damping with the ratio of 5%. The periodic live load at the middle of the beam is $F = 50 \sin (10t)$ kip. The stiffness of the beam can be calculated as $k = 48EI/L^3$.

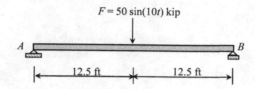

FIGURE 7.28   A Simply Supported Beam for Problem 7.13

Determine the following of the moment frame for the vibration in the horizontal direction:

(a) Single degree-of-freedom (SDOF) model
(b) Natural angular frequency, ω_n

    (c) Natural time period, τ_n

    (d) Frequency, f

    (e) Critical damping coefficient, c_c

    (f) Coefficient of damping, c

    (g) The differential equation of motion, i.e., the vibration equation (Damped forced vibration)

    (h) Steady-state displacement, X_p

    (i) Steady-state velocity, v_p

    (j) Steady-state acceleration, a_p

Appendix
Second-Order Linear Homogeneous Differential Equations with Constant Coefficients

An equation of the form, $y'' + ay' + by = 0$ can be solved by the method of undetermined coefficients where a solution of the form $y = Ce^{rx}$ is sought. Substitution of this solution gives

$$(r^2 + ar + b)Ce^{rx} = 0$$

and since Ce^{rx} cannot be zero, the characteristic equation must vanish or

$$r^2 + ar + b = 0$$

The roots of the characteristic equation are, $r_{1,2} = \dfrac{-a \pm \sqrt{a^2 - 4b}}{2}$

and can be real and distinct for $a^2 > 4b$, real and equal for $a^2 = 4b$, and complex for $a^2 < 4b$.

If $a^2 > 4b$, the solution is of the form (overdamped), $y = C_1 e^{r_1 x} + C_2 e^{r_2 x}$

If $a^2 = 4b$, the solution is of the form (critically damped), $y = (C_1 + C_2)e^{r_1 x}$

If $a^2 < 4b$, the solution is of the form (underdamped), $y = e^{\alpha x}(C_1 \cos \beta x + C_2 \sin \beta x)$

where: $\alpha = -\dfrac{a}{2}$, $\beta = \dfrac{\sqrt{4b - a^2}}{2}$

PROBLEM A.1

What is the solution of the differential equation, $y'' - y' - 12y = 0$?

Solution

Compare $r^2 - r - 12 = 0$ with $r^2 + ar + b = 0$, then the roots are

$$r_{1,2} = \frac{-a \pm \sqrt{a^2 - 4b}}{2} = \frac{1 \pm \sqrt{(-1)^2 - 4(-12)}}{2} = \frac{1 \pm 7}{2} = 4, -3$$

If $a^2 > 4b$, the solution is of the form (overdamped), $y = C_1 e^{4x} + C_2 e^{-3x}$

PROBLEM A.2

What is the solution of the differential equation, $y'' - 7y' = 0$?

Solution

Compare $r^2 - 7r = 0$ with $r^2 + ar + b = 0$, then the roots

$$r_{1,2} = \frac{-a \pm \sqrt{a^2 - 4b}}{2} = \frac{7 \pm \sqrt{(-7)^2 - 4(0)}}{2} = \frac{7 \pm 7}{2} = 0, 7$$

If $a^2 > 4b$, the solution is of the form (overdamped), $y = c_1 e^{0x} + c_2 e^{7x} = c_1 + c_2 e^{7x}$

PROBLEM A.3

What is the solution of the differential equation, $y'' - 5y = 0$?

Solution

Compare $r^2 - 5 = 0$ with $r^2 + ar + b = 0$, then the roots are

$$r_{1,2} = \frac{-a \pm \sqrt{a^2 - 4b}}{2} = \frac{0 \pm \sqrt{(-0)^2 - 4(-5)}}{2} = \frac{0 \pm 2\sqrt{5}}{2} = \sqrt{5}, -\sqrt{5}$$

If $a^2 > 4b$, the solution is of the form (overdamped), $y = c_1 e^{\sqrt{5}x} + c_2 e^{-\sqrt{5}x}$

PROBLEM A.4

What is the solution of the differential equation, $y'' - 3y' + 4y = 0$?

Solution

Compare $r^2 - 3r + 4 = 0$ with $r^2 + ar + b = 0$, then the roots are

$$r_{1,2} = \frac{-a \pm \sqrt{a^2 - 4b}}{2} = \frac{3 \pm \sqrt{(-3)^2 - 4(4)}}{2} = \frac{3 \pm i\sqrt{7}}{2} = \frac{3}{2} \pm \frac{i\sqrt{7}}{2}$$

If $a^2 < 4b$, the solution is of the form (underdamped), $y = e^{\alpha x} \left(C_1 \cos \beta x + C_2 \sin \beta x \right)$

where $\alpha = -\dfrac{a}{2} = \dfrac{3}{2}$, $\beta = \dfrac{\sqrt{4b-a^2}}{2} = \dfrac{\sqrt{7}}{2}$

$$y = e^{\frac{3}{2}x}\left(C_1 \cos \frac{\sqrt{7}}{2}x + C_2 \sin \frac{\sqrt{7}}{2}x\right)$$

PROBLEM A.5

What is the solution of the differential equation, $y'' - 8y' + 16y = 0$?

Solution

Compare $r^2 - 8r + 16 = 0$ with $r^2 + ar + b = 0$, then the roots are

$$r_{1,2} = \frac{-a \pm \sqrt{a^2 - 4b}}{2} = \frac{8 \pm \sqrt{(-8)^2 - 4(16)}}{2} = 4,4$$

If $a^2 = 4b$, the solution is of the form (critically damped), $y = (C_1 + C_2)e^{4x}$

Index

A

AASHTO, 248–250, 257, 261, 264
Acceleration, 2, 4, 6, 7, 10–24, 30–35, 37–41,
    43, 45–58, 63–69, 71–73, 75–77,
    79–82, 85–94, 120, 129, 130,
    132, 143, 144, 148, 158, 161–172,
    177, 178, 180–183, 189, 200, 201,
    224, 225, 229, 235–240, 243, 248,
    252, 258, 263, 266, 279, 280, 290,
    293, 300
Adhesion, 241, 242, 245, 246, 248, 263
Aerodynamic, 229–231, 240, 245–248, 259,
    263
Aero-elastic, 265
Amplitude, 272, 275, 277–280, 282, 283, 288,
    289, 294, 295, 299
Angular, 58, 137–142, 146, 156, 158–170,
    172, 173, 175, 177–185, 187, 189,
    198–201, 209, 213, 215–218,
    224–226, 265, 267–269, 271–275,
    278, 279, 282, 286, 290, 291, 294,
    296, 297, 299

B

Braking-efficiency, 245

C

Cantilever, 269, 297
Centrifugal, 52, 252–254, 256, 260
Centripetal, 54, 55, 252–255, 258
Centroid, 190, 192–197, 224
Conservation, 83, 93, 99, 116–120, 128–132,
    135, 138, 142, 216–218
Crane, 209, 210, 226
Crankshaft, 236
Curvature, 54–56, 67, 79, 81, 82, 252, 254, 256,
    257, 262, 263
Curve, 9, 10, 28–37, 41, 48, 49, 52–54, 56, 58,
    67–72, 76, 77, 79, 82, 122, 144, 145,
    164, 180, 252, 254–259, 262, 263,
    272, 286
Curvilinear, 48, 49, 52, 53, 57, 79, 137, 158,
    179, 201, 215, 254

D

Damping, 265, 266, 277, 278, 283–292,
    299–301, 303

Dashpot, 283, 287, 290, 299
Deceleration, 13, 17, 18, 22, 23, 71, 75, 100,
    114, 147, 163, 164, 230, 243, 245,
    248, 250, 251, 263
Deflection, 108, 266–268, 277, 289, 296
Deformation, 1, 148, 189, 205, 220, 232, 266, 281
Degree-of-freedom, 266, 287, 290, 299
Diaphragm, 278, 289
Displacement, 2, 9–16, 18–24, 27, 29–32,
    34–39, 41–47, 61, 72, 73, 76, 77, 90,
    91, 95–98, 113, 125, 143, 158–160,
    162–164, 167–169, 180, 181, 189,
    204, 207, 209, 266–268, 274,
    278–283, 286, 290, 292, 298, 300
Dissipation, 107, 117, 133, 211, 212, 265
Distance, 1, 2, 9, 13, 15, 18, 21, 32, 34, 38, 39,
    42, 43, 47, 56, 70–73, 75–79, 94,
    99–101, 103, 105, 107, 108, 110,
    115, 116, 119, 120, 137, 144, 148,
    149, 151, 157, 159, 189–191, 196,
    199, 200, 206, 215, 220, 221, 225,
    235, 240, 242, 243, 245–251, 255,
    259, 260, 263, 264, 268, 273, 276,
    284, 295, 296
Drag, 150, 230, 231, 237, 238, 262, 263
Dynamics, 1, 2, 4, 13, 189, 194, 229, 252, 266

E

Earthquake, 265
Eccentric, 296
Efficiency, 204, 211, 212, 225, 226, 230, 231,
    245, 246, 248, 263, 265
Electromagnetic, 84
Emission, 128
Energy, 75, 95, 98–101, 103–107, 109–123, 132,
    133, 142, 144, 146, 150–152, 189,
    201–207, 209, 211–214, 219, 221,
    225, 265, 294

F

Failure, 273, 278
Force, 1–4, 7, 9, 18–20, 26, 27, 41, 52, 54, 55, 60,
    61, 72, 83–90, 92–101, 104–112, 114,
    117, 122, 124–133, 138, 140–155,
    178, 189, 199, 200, 204, 205, 207,
    209, 213, 215–223, 226, 227, 229–
    241, 243, 245, 252–256, 258–260,
    262, 265, 266, 275–279, 281, 283,
    284, 286, 288, 289, 297–299

Printed in the United States
by Baker & Taylor Publisher Services